Cambridge Studies in Biological and Evolutionary Anthropology 38

Neanderthals and Modern Humans

Neanderthals and Modern Humans develops the theme of the close relationship between climate change, ecological change and biogeo-graphical patterns in humans during the Pleistocene. In particular, it challenges the view that Modern Human 'superiority' caused the extinction of the Neanderthals between 40 000 and 30 000 years ago. Clive Finlayson shows that to understand human evolution, the spread of humankind across the world and the extinction of archaic populations we must start off from a theoretical evolutionary ecology base and incorporate the important wider biogeographic patterns, including the role of tropical and temperate refugia. His proposal is that Neanderthals became extinct because their world changed faster than they could cope with, and that their relationship with the arriving Modern Humans, where they met, was subtle.

CLIVE FINLAYSON is Director, Museums and Heritage in the Government of Gibraltar, based at the Gibraltar Museum. He is also Professor in the Department of Anthropology at the University of Toronto. His research interests include Quaternary human–environmental patterns, the biogeography of hominids, and changing environments and faunal patterns in the Quaternary of southern Europe.

T0291446

Cambridge Studies in Biological and Evolutionary Anthropology

Series Editors

HUMAN ECOLOGY
C. G. Nicholas Mascie-Taylor, University of Cambridge
Michael A. Little, State University of New York, Binghamton
GENETICS
Kenneth M. Weiss, Pennsylvania State University
HUMAN EVOLUTION
Robert A. Foley, University of Cambridge
Nina G. Jablonski, California Academy of Science
PRIMATOLOGY
Karen B. Strier, University of Wisconsin, Madison

Neanderthals and Modern Humans

An Ecological and Evolutionary Perspective

CLIVE FINLAYSON

The Gibraltar Museum
and
The University of Toronto

CAMBRIDGE UNIVERSITY PRESS

CAMBRIDGE UNIVERSITY PRESS
Cambridge, New York, Melbourne, Madrid, Cape Town, Singapore, São Paulo, Delhi

Cambridge University Press
The Edinburgh Building, Cambridge CB2 8RU, UK

Published in the United States of America by Cambridge University Press, New York

www.cambridge.org
Information on this title: www.cambridge.org/9780521121002

First published 2004
Reprinted 2005
This digitally printed version 2009

A catalogue record for this publication is available from the British Library

Library of Congress Cataloguing in Publication data
Finlayson, Clive, 1955–
Neanderthals and modern humans : an ecological and evolutionary
perspective / Clive Finlayson.
        p.   cm. – (Cambridge studies in biological and evolutionary
anthropology ; 38)
Includes bibliographical references and index.
ISBN 0-521-82087-1
1. Neanderthals.   2. Human evolution.   3. Social evolution.
I. Title.   II. Series.
GN285.F55   2004
569.9 – dc21   2003055285

ISBN 978-0-521-82087-5 hardback
ISBN 978-0-521-12100-2 paperback

To Geraldine and Stewart

Contents

Preface

In 1848 a strange skull was discovered in Forbes' Quarry, Gibraltar, close to where I live. A second skull found eight years later in the Neander Valley, near Dusseldorf in Germany, gave a new hominid its name – the Neanderthal. This name, and its relation to an individual that lived close to the edge of its range, led to over a century of perception of the Neanderthals as a brutish people of northern Europe who survived, through thick and thin, the cold of the 'ice ages' until they were supplanted by the newly arrived and intelligent Modern Humans.

The image is still one that many regard as close to reality. Yet, paradoxically, the Neanderthals were intelligent people of mild climates. They evolved across the northern shores of the Mediterranean Sea and eastwards towards the Black and Caspian Seas. They ventured north only during mild climatic episodes and the unstable, cold and arid climate of late Pleistocene Europe eventually gave them the blow that sent them on the road to extinction. The Modern Humans hovered in the periphery and took advantage of the situations left vacant by the Neanderthals. This book is an attempt to redress the balance of over a century of misunderstanding.

Acknowledgements

I am grateful to the publishers, and in particular Tracey Sanderson, for the opportunity to publish this book and for their support throughout. The ideas put forward in this book were conceived after many discussions with friends and colleagues over a number of years. I am particularly indebted to my wife, Geraldine Finlayson, for her insightful discussions, ideas and support throughout. The ecological approach followed in this book stems from many years working in bird ecology. The ecological discussions have been particularly intense and fruitful with my friend and colleague Darren Fa.

I first ventured into the field of human evolution in 1990 when I became involved in the Gibraltar Caves Project. Two of its co-directors, Chris Stringer and Andy Currant of the Natural History Museum in London, have had a lot to do with my involvement and participation in this exciting field. I have been especially welcomed into the archaeological side of this subject, and have learnt vast amounts in the field, from the friendship and knowledge of Paco Giles of the Museo de El Puerto Santa María. I have spent many good times discussing and learning about the Palaeolithic from him and his team, especially Antonio Santiago Pérez, José María Gutierrez López and Esperanza Mata Almonte. I am also deeply indebted to my good friend and colleague Joaquin Rodriguez-Vidal for the brilliant way in which he has made me understand the geomorphology of the karstic landscapes that the Neanderthals lived in.

During the last five years in particular I have benefited from discussions with many colleagues, particularly during the two Calpe conferences organised in Gibraltar in 1998 and 2001: Emiliano Aguirre, Juan Luis Arsuaga, Javier Baena Preysler, Nick Barton, Ofer Bar-Yosef, Jacques Blondel, Eudald Carbonell, Miguel Cortés, Francesco d'Errico, Yolanda Fernández Jalvo, Rob Foley, Clive Gamble, Paul Goldberg, Marta Lahr, Richard MacPhail, Paul Mellars, Marina Mosquera, Paul Pettitt, Marcia Ponce de León, Robert Sala, Larry Sawchuk, Olga Soffer, Gerardo Vega Toscano, Erik Trinkaus, Manuel Vaquero, Joao Zilhao, Christoph Zollikofer.

1   *Human evolution in the Pleistocene*

The origins of humanity may be traced to the tropical African Pliocene, around 6 million years ago (Myr). Genetic evidence has for some time predicted the existence of a common ancestor to chimpanzees and humans around 5–6 Myr (Takahata & Satta, 1997; Gagneux & Varki, 2001). Recent discoveries of African fossils that are claimed to be close to this common ancestor have been dated to between 6 and 7 Myr (Brunet *et al.*, 2002).

From this point until the emergence of *Homo erectus* 1.9 Myr ago and its rapid subsequent range expansion (Aguirre & Carbonell, 2001), hominids were confined to sub-Saharan Africa. The estimated number of species that lived during this long period in the Pliocene varies among authors. If we follow a conservative approach (Klein, 1999) we observe a pattern of increasing hominid species richness from about 4.6 Myr with a peak between 1.9 and 1.6 Myr and a sharp decline thereafter (Fig. 1.1). The decline after 2 Myr ago is correlated with increasing climate instability.

The peak in diversity coincides with the first appearance in the fossil record of *H. erectus*. Recently this early African member of the genus *Homo* has been separated from contemporary Asian forms. The name *H. erectus* has been retained for the Asian forms and the name *H. ergaster* for the African (Klein, 1999). Recent evidence suggests, however, that the two significantly overlap in morphology and that they should form part of a geographically diverse species *H. erectus* (Asfaw *et al.*, 2002). I follow this latter classification here. Subsequent forms have been given specific status by different authorities although there is considerable uncertainty regarding the precise boundaries of each. The classification of fossils is fraught with difficulties as we shall see in Chapter 4. In this book I consider *H. erectus–H. sapiens* to be a single chronospecies (Cain, 1971) that has repeatedly produced divergent lineages through geographical isolation during the last 1.9 Myr. Some of the described forms are clearly temporal entities within the *H. erectus–H. sapiens* continuum. I include *H. heidelbergensis* and *H. helmei* in this category. Others are divergent lineages that have subsequently become extinct. The Neanderthals are the clearest example of such a divergent lineage and their relationship with mainstream *H. sapiens* will occupy much of this book. Until equivalent fossils are found in Africa it is probably best to regard the form *H. antecessor* from the Spanish site of Atapuerca (Carbonell *et al.*, 1995), and possibly also those of Ceprano

1

in Italy (Manzi *et al.*, 2001) in this latter category, i.e. a divergent lineage that became extinct.

The question of interbreeding between mainstream *H. sapiens* and divergent lineages when geographical or ecological barriers broke down will be addressed, with specific reference to Neanderthals and contemporary mainstream *H. sapiens*, in Chapter 7. The degree of genetic isolation of the constituent populations would be dependent on a range of factors at any point. These would include distance effects and physical, climatic and ecological barriers. Populations would become isolated at some points and a process of genetic divergence would ensue. Most often such a process would end with renewed contact among populations. At other scales, metapopulations in different regions would become isolated from each other. Gene flow would continue within but not between regions. At even larger spatial scales entire regions would occasionally

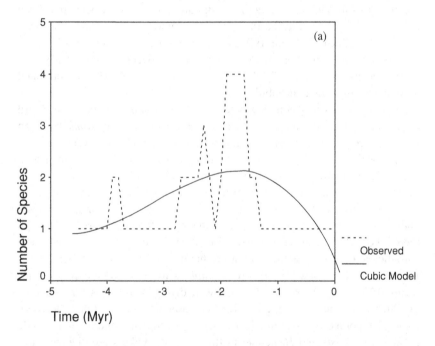

Figure 1.1. (a) Number of hominid species during the last 5 Myr using a conservative number of species. A cubic model best fits the observed pattern: $y = 0.2328 - 2.5022x - 0.9973x^2 - 0.1059x^3$; $R^2 = 0.293$; $P = 0.002$. (b) Decline in hominid species in the last 2 Myr. A cubic model best fits the pattern: $y = 0.8187 - 2.5122x - 5.6201x^2 - 3.1246x^3$; $R^2 = 0.923$; $P < 0.0001$. (c) Relationship between number of hominid species and climate variability (coefficient of variation of temperature) in the last 2 Myr. The pattern is best described by a quadratic model: $y = 10.9797 - 1.9269x + 0.914x^2$; $R^2 = 0.366$; $P = 0.033$.

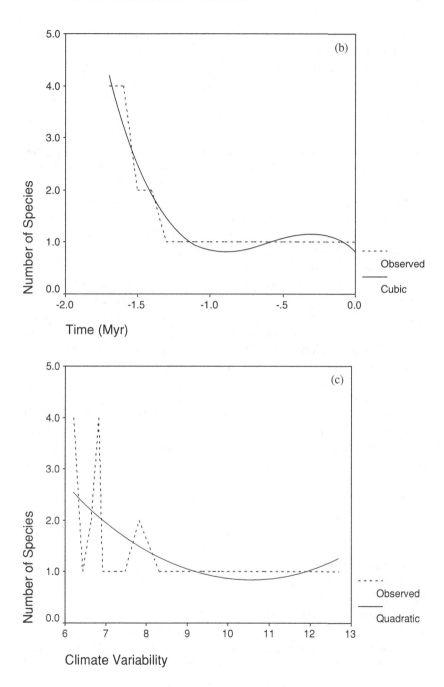

Figure 1.1. (*cont.*)

become isolated from others. I have introduced scale here and it is an issue that is central to understanding ecology (Levin, 1992) and will appear frequently in this book. In this case we can see how small-scale population isolation events would be expected to be frequent relative to regional events involving many populations.

Populations most distant from each other would be expected to be genetically most distinct but linked to each other by intermediate forms. Where isolation of extreme populations was long, populations at the extremes of the range may have diverged to the extent that they subsequently behaved as good species. In the case of Pleistocene *Homo*, geographical comparisons have to be made among contemporary forms. As we are studying phenomena through time, it is also important that geographical patterns from different time periods are not merged. It is common, for example, to find generalised distribution maps of Neanderthal geographical range in the literature (e.g. Stringer & Gamble, 1993). These should only be regarded as maps of the extremes of the range reached according to currently available evidence. In reality the Neanderthal range, as that of other forms of *Homo* and indeed all other animals, shifted, expanded and contracted through time and it is these range changes that are likely to be most informative about Neanderthal behaviour, as we shall see in Chapter 3 (Fig. 1.2). If we follow this approach, bearing in mind the limitations of the available data, we observe a changing pattern of global distribution of *Homo* in the Pleistocene.

There are two apparently contrasting models that, as we shall see in this book, are in effect extremes of a continuum. Much of the debate that has raged in the last two decades in this respect has been due to differences in the under-standing of the evolutionary process and confusion with taxonomic techniques, particularly cladistics. I will start with a brief statement of the two contrasting models.

On the one hand, we have the multiregional model that has been championed by Wolpoff and his school (Wolpoff, 1989). According to this model *H. erectus–H. sapiens* is a single species (hence *H. sapiens*). The variations that are observed among fossils simply reflect natural variation as the species has evolved through time. As populations became isolated, so geographical variations arose between them just as they do in most widely distributed organisms. According to this model and its variants, genetic barriers between the populations were never severe enough to cause speciation. Thus present-day human populations reflect a combination of regional variation that dates back to the earliest colonisations and relatively continuous gene flow among the populations. The intensity and frequency of gene flow would be greatest among neighbouring populations and lowest among those geographically most distant.

Figure 1.2. Maximum limits (grey area) of the Neanderthal geographical range in Europe and western and central Asia. Bioclimate boundaries as in Figure 5.3.

On the other hand, we have the 'Out-of-Africa 2' model that has been associated most strongly with Stringer (Stringer & Andrews, 1988). According to this model all natural variation that existed among populations of *Homo* was removed very rapidly after 100 000 years (kyr) ago by the geographical expansion of 'Modern Humans' that evolved somewhere in eastern or north-eastern Africa. As these 'Modern Humans' spread out of Africa they replaced all existing populations of *Homo* across Africa and Eurasia. These 'Archaic' African and Eurasian populations had evolved regionally after an earlier 'Out-of-Africa 1' expansion of *H. ergaster* around 1.9–1.8 Myr ago. The model, in its current form, does not negate the possibility of interbreeding among 'Modern' and 'Archaic' forms on contact but it does assume that no 'Archaic' genes persisted into present-day populations.

These ideas may seem very different and irreconcilable but in reality this is not the case. To a large extent the two views reflect a different understanding of the evolutionary process. The multiregional model follows the neo-Darwinian school that sees evolution proceeding through small, cumulative, changes within a species. The macro-evolutionary changes observed in the fossil record are simply the accumulation of many micro-evolutionary changes. Thus *H. ergaster/ erectus* gradually evolves into *H. sapiens*. Any division of the lineage into species is of necessity arbitrary. This interpretation is correct. New species arise when populations of a species are isolated from each other sufficiently so that when they secondarily meet they do not hybridise to an extent that the two populations eventually become one (Cain, 1971). Thus the multiregional model, whether correct or not, is consistent with neo-Darwinian evolutionary theory.

In the 1970s and subsequently, Gould & Eldredge (1977) proposed a different evolutionary process. Coming from a palaeontological background these authors had difficulty in understanding how the major steps (such as apparently sudden adaptive radiations) observed in the fossil record could arise through the accumulation of many micro-evolutionary changes. They saw the evolutionary process as a series of major steps punctuated by long periods of stasis during which species shifted their adaptive positions within defined parameters but without significant speciation taking place. No clear mechanism has been satisfactorily defended for such a process. At about the same time a new taxonomic methodology was being developed. Cladistics was seen as a quantitative and objective method of classifying species that significantly improved on existing phylogenetic procedures. By measuring a suite of variables (usually metric), taxonomists were able to separate those that were common to a lineage from those that were specific to a lineage. Whenever such specific differences were observed in a form it was given specific status. Thus, if we understand evolution as being driven by speciation events we move to a situation in which, as new species arise (or are defined cladistically which is not the same thing!), the ancestral ones *de facto* cease to exist. We can now begin to understand why the replacement school (that relies heavily on cladistics) has difficulty in accepting a *H. ergaster/erectus* – *H. sapiens* continuum. Instead, it sees every new fossil that is discovered and has features specific to its lineage as a new species.

In reality the evolutionary process proceeds in two ways: through the gradual accumulation of small changes within a species and through the formation of new species, in vertebrates at least in geographical isolation, through a process known as allopatric speciation. Recent studies seem to be providing evidence for speciation within a common geographical area through the combination of ecological and behavioural differences within a population (sympatric and parapatric speciation) (Maynard Smith, 1966; Rice & Hostert, 1993; Gavrilets *et al.*, 1998; Dieckmann & Doebeli, 1999; Kondrashov & Kondrashov, 1999; Tregenza & Butlin, 1999; Danley *et al.*, 2000; Filchak *et al.*, 2000; Johannesson, 2001; Porter & Johnson, 2002).

There are inconsistencies in the 'Out-of-Africa 2' model that are attributable to not giving importance to gradual micro-evolutionary processes. Thus, if 'modern humans' emerged in Africa they must have done so, according to this view, via a speciation event. An alternative, that is more parsimonious and equally valid, is that 'modern humans' evolved differences gradually over the last 2 Myr from the ancestor of the hominids that spread to other parts of Africa and into Eurasia. To accept this position would imply acceptance of regional continuity in that part of Africa at least. It is these humans that I term mainstream *H. sapiens*, the 'Moderns', in this book.

The next difficulty arises in the definition of species that, as we have seen already, is fraught with difficulties because we are unable to apply the bio-logical species concept to fossils. It is presumably one reason why palaeo-anthropologists and archaeologists are so hotly debating the Lagar Velho fossil from Portugal that is purported to be a Neanderthal–Modern hybrid (Duarte *et al.*, 1999; Zilhao & Trinkaus, 2002). There is no doubt that the Neanderthals at least were a separate lineage in human evolution. Using cladistics that makes them a separate species. This need not be the case. The Neanderthals may have embarked on a separate evolutionary course from mainstream *H. sapiens* but the degree and time of isolation when the two lineages re-met in the Middle East and later in Europe would have determined whether or not they were a good species. It is largely a question of detail that has little bearing on the study of the two populations other than on the question of interbreeding which will be very hard to resolve in any case. For these reasons I will develop the arguments in this book along the lines of populations as this will be a more productive approach. I will utilise nomenclature only in so far as it aids the reader. Nothing more should be made of the use of particular names.

The multiregional model, on the other hand, does not appear to attach impor-tance to the geographical replacement of one population by another. Yet, there are many examples in the literature of the spread of populations and species, which is a part of the dynamics of the natural world. It seems unlikely that, in the history of the genus *Homo*, there should only have been a single successful 'Out-of-Africa' expansion. Implicit in the multiregional model is the failure of any subsequent population expansion other than through genetic assimilation. In the case of the Moderns and the Neanderthals in Europe, it would seem that current evidence clearly indicates the replacement of the Neanderthals by the Moderns. It is a different expectation, and to my mind an unrealistic one, to as-sume that such replacement need have been worldwide. In any case, as we shall see later, the colonisation of Europe by Moderns need not have been strictly a replacement, if by that we mean an active displacement of Neanderthals by the new arrivals.

The thrust of this book will, I hope, shed a new light on the processes and the mechanisms that have marked the course of human evolution. The basis of the argument has been marked out by Finlayson *et al.* (2000a) who have adopted a biogeographical approach that sets off from an evolutionary ecology stance. According to this view the growth of Modern Human populations and the decline and extinction of the Neanderthals were independent, climate-linked, events. Modern superiority, leading to the disappearance of the Neanderthals through competition, was considered implausible. The initial colonisation of the world by Moderns was related to a coincidence of climatic and historical events that favoured a population that was adapted to the exploitation of plains mammalian herbivores. The geography of the northern hemisphere and climate-induced vegetation changes coincided to make the colonisation successful.

One of the criticisms of the contrasting models set out above (especially the 'Out-of-Africa 2') is that a mechanism has not been put forward to explain the model. Equally, testable predictions have not been generated. In this book I will develop an ecological and evolutionary perspective that attempts to understand human evolution through that of its constituent populations. Climate is seen as a central element that has been critical in human evolution, not necessarily directly as some have postulated (Ruff, 1994; Holliday, 1997a, b) but rather through its effects on the distribution and abundance of plants and animals. I highlight, in particular, the increasing climatic instability during the Pleistocene as a critical factor that has been largely ignored (but see Potts, 1996a, b; 1998), although in my view a new mechanism of 'variability selection' is not required, as I will explain later. Running in parallel with the climatic and ecological vicissitudes of the Pleistocene, humans have evolved mechanisms to deal better with these uncertainties. These mechanisms have, in the end, permitted the colonisation of the entire planet.

2 *Biogeographical patterns*

The distribution and abundance of plants and animals during the Quaternary is of great interest in the understanding of the pattern for any particular species. In our case it is fundamental to understanding the way in which humans were distributed at different times during the Quaternary.

It is important to start our discussion at the macro-ecological scale. The broad biogeographic picture will give us important insights at the scale which is most relevant to our study. We will zoom into lower spatio-temporal scales in later chapters where it is relevant to the discussion. I will not spend time discussing well-established biogeographic patterns that I do not regard to be especially relevant to this book. I am more concerned with the distribution and shifts in distribution of environments that would have influenced human distribution and I will confine my discussion largely to the Eurasian and African land masses which is where the main events took place.

Vegetation structure

In this book I will place particular emphasis on vegetation structure, that is the three-dimensional arrangement of plants in space. The reason for this is that I consider that vegetation structure will have played a major role in the distribution of humans, as it does for most animals (Bell *et al.*, 1991). Vegetation structure would have been particularly important in determining the types of potential prey available to humans and also in making prey visible and accessible. Part of the reason why forests were among the last habitats to be colonised by humans (Gamble, 1993) must have had to do with prey visibility and accessibility as well as density.

We may describe vegetation structure according to the distribution of plants on the ground layer (forbs and grasses), the shrub layer and the tree layer (Kent & Coker, 1992). Even though the species composition will vary significantly between regions of the world, vegetation structure shows similarities. For the purpose of this book I will cluster habitats by vegetation structure into the following blocks.

9

Forests

There is a predominance of trees at high density with a dense canopy. Variants include tropical and equatorial rainforests, where the canopy is very high, and temperate broad-leaved forests.

Shrublands

There is a predominance of shrubs with the virtual absence of trees. Today, characteristic examples are the Mediterranean shrublands, known by specific names in different parts of the world (e.g. matorral, chaparral; Cody, 1974).

Open habitats

These are characterised by the absence (or presence in low density as in wooded savannah) of trees and shrubs and a predominance of grasses, forbs, mosses or lichens, or by the total absence of vegetation in patches. Savannahs, steppe and tundra cluster under this definition.

Deserts

Deserts occupy large areas of the planet and are characterised by the virtual absence of vegetation on account of low and irregular rainfall (Cox & Moore, 1985). There are sandy, rocky and ice deserts. Deserts are therefore a separate category of habitat that cannot be described adequately by vegetation structure other than as extreme open habitats. For the purpose of this book I will consider deserts to be a separate category. In human terms deserts have played a major role as barriers to dispersal. Human adaptations to deserts are extreme developments of adaptations to open habitats.

Rocky habitats

These are areas with a minimal vegetation development and a preponderance of a rocky substrate that, like deserts, may be considered extreme cases of open habitats. Unlike deserts they are usually localised at the landscape and regional scales. Two types of rocky habitats have been particularly important to humans. Where the inclination of the land is vertical, or nearly so, rocky habitats are described as cliffs. Cliffs have attracted humans as areas for shelter or where specialised fauna (e.g. ibexes *Capra* spp.) are concentrated. Within rocky areas, especially in karstic environments, are cavities. These have been traditionally used by humans as shelters.

Wetlands

Lakes, marshes, alluvial plains, rivers and estuaries and deltas are special habitats. They are usually localised on a regional scale. Their main characteristic is the presence of standing water (usually fresh or brackish). Margins will often be vegetated with grasses, reeds and shrubs. Wetlands, depending on climate, may be seasonal. They attract concentrations of animals at specific times of the year and are additionally sources of animals not found in other habitats, especially fish. Wetlands have been extremely important to humans throughout the Quaternary (Nicholas, 1998).

The sea

Human exploitation of the open ocean is a recent phenomenon (Gamble, 1993; Fernández-Armesto, 2000). The products of the sea have, however, been exploited by humans in coastal areas since, at least, the last interglacial (Balter, 2001). Like deserts, the sea has often played a major role as a barrier to human dispersal even though this has not always been the case, the colonisation of Australia before 50 kyr ago being a case in point (Thorne *et al.*, 1999; Bowler *et al.*, 2003).

Mosaics: transitional and edge habitats and heterogeneous landscapes

The habitat categories that I have so far described will be those that I will be using throughout this book. They are habitats from the human perspective. Where these habitats meet there may be sharp discontinuities between one and the other. These edge areas or ecotones are areas of high diversity (Kerr & Packer, 1997). These could occur, for example, where wetland and forest come into contact or where plains or cliffs come into contact with the sea in coastal areas or indeed where forest and open habitats are close to each other.

After a perturbation an area may experience a succession of habitats over a period of time (Bazzaz, 1996). The classic example is the regeneration of woodland after a fire. Depending on the point in time at which we look at an area we may observe it in transition. This is not unusual and it is my contention that such transitional situations were the rule at particularly critical stages in the Quaternary. The abrupt climatic changes that have been recorded at the scale of decades and centuries with the consequent rapid alterations to the vegetation (Chapter 6) meant that large areas of the world would have had transitional habitats for long periods. Given that the climatic peaks occupied a small proportion of the Quaternary (Lambeck *et al.*, 2002a, b) and that, even these peaks were often highly variable, we have to accept that large areas of the planet that were occupied by humans during the Quaternary would have been dynamic in habitat features at the scale of human generations.

Finally, where spatial discontinuities exist in critical variables at the landscape or regional scales we find habitat mosaics rather than uniform blankets of single habitat (Forman, 1995). Such mosaics are especially common today as humans continue to modify the environment but they would have always existed. Such mosaics would, like edge and transitional habitats, have offered opportunities for humans to exploit the natural diversity within.

Altitude

I do not consider mountains as a specific habitat category in this book. The habitats described so far may be found at high altitude and their extent would have varied in most cases in response to climate changes in a similar manner to latitude (MacArthur, 1984). The highest mountains, however, acted as physical barriers to human dispersal, especially in the coldest moments when they were virtually impenetrable. In Eurasia, the Himalayas continue to be a barrier even today. The belt of mountains stretching from the Iberian Peninsula and the Maghreb in the west to the Himalayas in the east was critical in human evolution (Finlayson *et al.*, 2000a). Large changes in altitude over short distances produced landscape mosaics with high local biodiversity as happens today (Cody, 1986). This was, in my opinion, critical to the evolution of the Neanderthals. In contrast, the generally low-lying and topographically homogeneous Eurasian Plain, stretching from Britain to the Bering Strait, was only fully colonised by humans very late in the Pleistocene (Chapter 7).

Habitat changes in the Quaternary

The climatic oscillations of the Quaternary, through changes in temperature and rainfall, produced many large-scale changes in the geographical distribution and the extent of a number of the habitats described above. These changes are summarised below.

Forests

Tropical and equatorial rainforests contracted their range significantly during arid events that were associated with increasing cold (Lezine *et al*, 1995; Colinvaux *et al.*, 1996; Dam *et al.*, 2001) and expanded their range during wet periods that were associated with warm events. Temperate broad-leaved forests expanded from their European strongholds eastwards during warm and wet events and contracted westwards during cold and arid ones (Chapter 6; Zagwijn, 1992). These forests expanded the northern edge of the range in warm events reaching as far north as Scandinavia. The northern edge of the range of these forests contracted in cold and arid events (Chapter 6; Zagwijn, 1992). The expansion on the southern edge of the range was limited by the Mediterranean Sea. Boreal coniferous forests shifted their range north and south in response to warming and cooling (van Andel & Tzedakis, 1996). In the Mediterranean, montane coniferous forests shifted their range up and down mountains in a similar manner whereas the thermophillous Mediterranean pines reached their maximum extent in interglacials (Finlayson, 1999).

Shrublands

Shrublands would have fluctuated in area as transitional habitats, such as forests, gave way to open habitats and vice versa. In the Mediterranean Basin, Mediterranean shrubs persisted throughout the Quaternary. Their range would have contracted at the expense of forest in warm and wet periods and at the expense of steppe in cold and arid ones (Carrión *et al.*, 2000). Shrublands would therefore have occupied large expanses of the Mediterranean Basin at different times in the Quaternary. Although their extent has increased as a result of human action through deforestation the Mediterranean shrublands would appear to have a long evolutionary history (Blondel & Aronson, 1999). To the north and south of the Mediterranean the more extreme boreal and tropical conditions are likely to have led to more rapid and abrupt changes from forest to open habitats and back. The intermediate position of the Mediterranean lands would have made them best suited for the development of shrubland communities and habitats.

Open habitats and deserts

In Africa, savannahs and grasslands expanded at the expense of rainforest during cold and arid periods and at the expense of desert during warm and wet periods (Chapter 6; Dupont *et al.*, 2000) and vice versa. The maximum extent of the

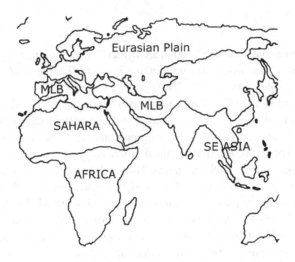

Figure 2.1. Distribution of main habitat and topographic blocks referred to in this book. MLB: mid-latitude belt.

Sahara would have been reached during most arid moments (Swezey, 2001) when it would have been a barrier to human dispersal (Marks, 1992; Lahr & Foley, 1994). In the wettest events, on the other hand, the Sahara was virtually taken over by grasslands and savannah. During such times its effect as a barrier to dispersal would have been insignificant. The development of grasslands in South-east Asia would have followed a similar pattern except that their extent would never have been as great as in Africa (Dam *et al.*, 2001).

In Eurasia the expansion of steppe westwards occurred during arid events (van Andel & Tzedakis, 1996). Its western limits receded significantly during warm and wet events and the expansion of forest. At their maximum extent, steppes covered much of the central tablelands (mesetas) of the Iberian Peninsula. Tundra expanded south and west during glacials. The ice deserts expanded southwards, reaching their maximum extent during the coldest and wettest glacials. In the Mediterranean, mountain glaciers responded in a similar manner.

Contrasting equatorial, tropical and sub-tropical Africa, the intermediate mountainous belt and the northern plains

In terms of habitat and topographical characteristics that would have been significant to humans we may divide Africa and Eurasia into three major blocks (Figure 2.1).

Tropical and sub-tropical Africa

This is essentially the geographical area from within which hominids, including humans, originated and dispersed. In habitat terms there has been a dynamic expansion and contraction of rainforest, savannah, grassland and desert throughout the Quaternary (deMenocal 1995; Dupont *et al.*, 2000; Swezey, 2001). The reduction in rainforest at the expense of open habitats has been proposed as a major factor in human evolution (Foley, 1987; Foley & Lee, 1989; Kingston *et al.*, 1994). The contraction of the Sahara has been proposed as a major factor permitting the dispersal of tropical African animals, including humans, northwards. The combination of plains and heterogeneous landscapes, particularly along the Rift Valley, would have produced ample opportunities for ecological diversification and allopatric isolation among hominids (O'Brien & Peters, 1999).

The intermediate mountainous belt

This is the belt that stretches from Iberia and the Maghreb in the west to the Himalayas in the east. Any Eurasian population to the north must have dispersed from this area. Its southerly position within the Eurasian landmass made this belt suitable for permanent or semi-permanent human occupation (Finlayson *et al.*, 2000a). These lands, because of their latitude, would have been less affected by the severity of the glaciations than the plains immediately to the north. The west would have been especially suitable on account of the oceanic influence of the Atlantic. This belt therefore provided a number of refugia for many species, including humans, during the glaciations (Hewitt, 1999).

Different parts of this wide longitudinal area would have offered different opportunities on account of their characteristics (Finlayson, 2003). The Iberian Peninsula in the west would have been the major refuge for European human populations being largest in area of the Mediterranean peninsulas and because of the milder oceanic climate, especially along its coasts. The interior tablelands would have experienced more severe climatic situations. The Strait of Gibraltar, immediately to the south, would have been open throughout the Quaternary so that any human movement between Europe and Africa, if it took place, would have involved a short sea crossing that, at its best, would have involved a series of island hopping events with the longest sea crossing being of the order of 7km (Alimen, 1975; Giles Pacheco & Santiago Pérez, 1987). The Maghreb, on the other side of the Strait of Gibraltar, would have been isolated by the sea to the north and by the Sahara to the south (in cold and arid moments). Human populations living in the Maghreb would have had the possibility of contact with populations further south when the Saharan barrier broke down. The question remains whether or not these North African humans had contact with the Iberian populations (Simmons & Smith, 1991; Hublin, 1992).

Further east, Italy and the Balkans would also have been refuges for humans during glaciations but their greater continentality and reduced surface area when compared to Iberia would have made them less important. The Mediterranean Sea would have been a barrier to contact with North Africa.

The next major refugium would therefore have been Turkey and the Levant. The high mountains of Turkey would have limited human populations to the narrow coastal strip during glaciations although reduced sea levels then would have widened the area significantly (Finlayson *et al.*, 2000a). The Levant would have contrasted with Iberia and the Maghreb in that the absence of a sea barrier would have brought Eurasian humans into more frequent contact with north-eastern Africans. As with the Maghreb, Levant populations would have had the opportunity of contact with Africans when the Sahara barrier broke down although there may have been a regular contact thoughout via the Nile Valley (van Peer, 1998).

Our knowledge further east is less complete but Crimea, areas around the Caucasus and even further east have Mediterranean bioclimates even today and are likely to have been refugia for Eurasian hominids (Finlayson *et al.*, 2000a).

Finally, though not a part of this belt, the temperate regions of southern Africa reveal similar characteristics to those of northern Africa. The absence of the Sahara or of a sea barrier would, however, have permitted a greater degree of contact with tropical African populations than would have been the case with the Maghreb, the Levant or the northern shore of the Mediterranean.

The Great Eurasian Plain

The low-lying land that stretches almost without interruption from Britain to the Bering Strait was covered in broadleaved forest in the west, steppe in the east and boreal coniferous forest in the north. To the north were the glaciers and ice sheets. Throughout the Quaternary these four elements grew and shrank at the expense of the other depending on temperature and rainfall (van Andel & Tzedakis, 1996). The milder, western, part was colonised by humans at a relatively early stage, perhaps before half-a-million years ago (Stringer & Gamble, 1993; Klein, 1999). The open plains of eastern Europe and western and central Asia were, on the other hand, colonised late. To my knowledge no human ventured into these landscapes before the Moderns carrying an Upper Palaeolithic tool kit at around 45–40 kyr BP (Bar-Yosef, 2000). Once colonised, the open plains provided little physical resistance to dispersal. With the spread of steppe westwards during the latter part of Oxygen Isotope Stage (OIS) 3 around

45–40 kyr came a steppe-adapted fauna and with them came the Upper Palaeolithic people with an Aurignacian tool kit (Finlayson, 2003).

South, South-east and East Asia

The southern and eastern parts of Asia, below the Himalayan barrier, are distinct
enough to merit separate treatment. The key feature of this large area of land
is that hominids dispersing from Africa would have found it relatively easy
to penetrate in comparison to the areas to the north of the Levant. Access
would have been facilitated further during low sea levels when the coastal
belt widened and access may then even have been gained from the Horn of
Africa (Lahr & Foley, 1994). In addition, hominids dispersing in this direction
would have kept to low latitudes and therefore similar seasonal, climatic and
environmental regimes (Finlayson *et al.*, 2000a). Thus eastern and south-eastern
Asia were colonised at a very early stage over 1 Myr ago (Klein, 1999). A fairly
constant flow of genes between these Asian and their African counterparts
would thus be expected with fewer interruptions due to climatic shifts than on
the main Eurasian landmass or even due to the Saharan expansion. The main
factor countering such flow would be the distance effect at the extremes. I have
proposed elsewhere (Finlayson *et al.*, 2000a) that tropical South-east Asia, once
colonised, probably had continuity of human occupation until today. The long
persistence of *Homo erectus* to perhaps as recently as 25 kyr (Swisher *et al.*,
1996) would support this prediction. The question of what happened to these
populations is uncertain. Tropical South-east Asia therefore became the second
most important glacial refugium for hominids after tropical Africa itself and,
as with the Maghreb, Levant, southern Africa and Europe, temperate zones
of eastern Asia would have been colonised from here during mild conditions.
The difference with tropical Africa is that hominids could spread northwards
more easily given the absence of the Saharan barrier. I therefore consider southeastern and eastern Asia to have constituted a separate system from the African–
Eurasian system in terms of hominid biogeography once the latter was settled.
Gene flow between the two would have been least at the extremes allowing for
a certain degree of separate evolution. The situation of India, which is largely
unknown, is of great interest in this respect.

The periphery

Two large land masses that lie on the periphery of the Eurasian–African systems
so far described, Australia and the Americas, were colonised late. Predictably,
Australia was colonised much earlier than the Americas (Thorne *et al.*, 1999;

Bowler *et al.*, 2003). This would be in keeping with the view proposed above that the Asian tropics were, despite their distance from Africa, ecologically easier to enter than the temperate north. Once in South-east Asia it was a question of time before the sea barrier to Australia was crossed. Once achieved, as with the Eurasian plains, a rapid expansion would be expected.

The key factor for the colonisation of the Americas would have also been the arrival in eastern Siberia. This is the easternmost extension of the Eurasian Plain which, as we have seen, was colonised late (around 42 kyr) and by Moderns (Brantingham *et al.*, 2001). As with these plains and Australia a rapid expansion would be expected once the barrier of the Bering Sea was surmounted.

Mammalian herbivores

Although humans are omnivorous animals, and are likely to have been throughout their evolutionary history, their relationship with herbivorous mammals would have been a major factor that permitted the range expansion outside the African tropics. Mammal meat would have been particularly important in more northerly latitudes where many of the alternative food sources available in the tropics would not have been found. It is therefore important to have an understanding of the distribution of mammalian herbivores (Table 2.1).

Proboscideans (Order Proboscidea)

Proboscideans would have been a feature of the many landscapes inhabited by humans in the Quaternary on most continents. In Africa, Proboscideans would have included a late representative (*Deinotherium*) of the Deinotheriidae, a Miocene family that survived until just over 1 Myr in east Africa (Shoshani *et al.*, 1996). These were specialised animals adapted to processing soft foliage that had become rare in the early Pleistocene, apparently due to the spread of grasslands and competition from Elephantoids in the Pliocene. The dominant Proboscideans of Quaternary Africa were the Elephantidae. Three major genera evolved in Africa. *Loxodonta*, the genus of the present-day African Elephant *L. africana*, evolved in Africa at the end of the Miocene around 6.2–5.6 Myr (Kalb *et al.*, 1996) and remained essentially an African lineage (Todd & Roth, 1996). *Loxodonta atlantica* persisted into the mid-Pleistocene in Africa alongside *L. africana*. Earlier African Plio-Pleistocene forms were *L. adaurora* and *L. exoptata* (Kalb *et al.*, 1996). The other two African genera are thought to have shared a more recent common ancestor than each did with *Loxodonta* (Kalb *et al.*, 1996). *Mammuthus* appears in South Africa in the early Pliocene

Table 2.1. *Main mammalian herbivore categories, their geographical distribution and habitat occupation in the Pleistocene and the degree to which they were exploited by humans*

Herbivore type	Size class	Africa	MLB	Eurasian Plain	SE Asia	Open	Intermediate	Closed	Rocky	Wetland	Hunted	Domesticated
a. Elephant	L	++	++	++	++	++	++	+	-	+	+	-
b. Horse[1]	M	++	+	++	-	++	+	-	-	+	++	++
c. Tapir	M	+	-	-	++	-	-	++	-	-	-	-
d. Rhinoceros	L	++	++	++	++	+	++	+	-	+	++	-
e. Pig[2]	S/M	++	++	+	+	+	++	+	-	++	++	++
f. Hippopotamus	L	++	+	+	++	-	+	-	-	++	+	-
g. Camel[3]	M/L	-	++	+	-	++	-	-	-	-	-	+
h. Mouse Deer	S	-	-	-	++	-	-	++	-	-	-	-
i. Giraffe	L	++	-	-	-	-	++	-	-	+	+	-
j. Deer[4]	M	-	++	++	++	++	++	++	-	++	++	+
k. Alcephalines	M	++	-	-	-	++	+	-	-	+	+	-
l. Antelopes	S/M	++	+	+	-	++	++	-	-	+	++	-
m. Bovids[5]	M/L	++	++	++	++	++	++	+	-	++	++	++
n. Caprids[6]	M	+	++	-	-	-	+	-	++	-	++	++
o. Duikers	S	++	-	-	-	-	-	++	-	-	-	-
p. Other	M	++	-	-	-	+	++	-	-	++	+	-

Type categories: a. all Proboscidea; Perissodactyla: b. Equidae; c. Tapiridae; d.Rhinocerotidae; Artiodactyla: e. Suidae; f. Hippopotamidae; g. Camelidae; h. Tragulidae; i. Giraffidae; j. Cervidae; Bovidae: k. Alcelaphinae; l. Antelopinae; l. Antelopinae; m. Bovinae; n. Caprinae; o. Cephalophinae; p. Hippotraginae, Peleinae, Aepycerotinae, Reduncinae.

Domesticated species (from Diamond, 1997): 1, horse, donkey; 2, pig; 3, dromedary, bactrian camel, alpaca; 4, reindeer; 5, cattle, water buffalo, yak, Bali cattle, mithan; 6, sheep, goat.

MLB: mid-latitude belt.

(*M. subplanifrons*). This species and *M. africanavus* of Central and North Africa (the ancestor of *M. primigenius*) became extinct in Africa by the end of the Pliocene and beginning of the Pleistocene (Haynes, 1991). The third genus, *Elephas*, first appears as *E. ekorensis* in Kenya around 3.75 Myr (Kalb *et al.*, 1996). The genus, subsequently represented in Africa by *E. recki* and *E. iolensis*, became extinct in that continent during the last glaciation (Todd & Roth, 1996).

The Mammutidae are absent from Plio-Pleistocene Africa but they persisted in Eurasia where they arrived as African immigrants in the Mid – Upper Miocene (Tobien, 1996). *Mammut borsoni* appears in the Upper Miocene of China. The species is common in Europe including the Mediterranean until the early Pleistocene (Tobien, 1996). It is therefore possible that early hominids dispersing into Asia encountered this species but it is less likely that they did so in Europe. The Stegodontidae emerged in Africa in the early Messinian (around 6.5 Myr) but had disappeared by the late Miocene (around 3 Myr). Migrations into Asia took place during lowered sea levels at the end of the Miocene and the family thrived in Asia (Kalb *et al.*, 1996). The genus *Stegodon* is present in numerous localities in China, south and east Asia from the late Miocene (around 5.5–5.3 Myr) where it persisted well into the Pleistocene (Kalb *et al.*, 1996). Among the Elephantidae, there was a first African dispersal of *Elephas* into Eurasia in the middle Pliocene (around 3.5 Myr) and a second, of *Elephas* and *Mammuthus*, in the late Pliocene (around 2.5 Myr) (Todd & Roth, 1996). The first dispersal, into Asia, led to the evolution of the present-day Asian Elephant, *Elephas maximus*, while the second led to *E. antiquus* and *E. falconeri* in the middle Pleistocene, which became extinct during the last glaciation. Mammoths (*Mammuthus*) dispersed rapidly across Eurasia and North America in the late Pliocene. These evolved towards progressively more open landscape forms: southern elephant *M. meridionalis* in wooded steppe; steppe mammoth *M. trogontheri* in cool, dry steppe with scattered trees; and woolly mammoth *M. primigenius* in cold, treeless environments. The mammoth lineage evolved in response to the cooling of the Pleistocene climate and the expansion of open habitats, although competition from middle Pleistocene *Elephas* in the warmer, forested, areas of Eurasia may have been a contributing factor (Lister, 1996). The last mammoths died out in Siberia around 10 kyr (Martin & Klein, 1984) with an isolated population on Wrangel Island in the Arctic Ocean persisting until 4 kyr (Vartanyan *et al.*, 1993).

North America saw the late survival of Proboscidean groups that had become extinct much earlier in Africa and Eurasia. There were two immigrations into North America across the Bering Strait, one in the middle Miocene (around 15.5 Myr) that involved Gomphotheriidae, Ambelodontidae and Mammutidae, and a second in the Pleistocene that involved the Elephantidae (*M. meridionalis* and *M. primigenius*) (Saunders, 1996). The mammoth dispersal may have involved an early entry of *M. meridionalis* around 2.0–1.9 Myr and a second, of *M. primigenius*, just before the start of the last glaciation (Dudley, 1996). Of these radiations and subsequent evolution, *Cuvieronius tropicus* (Gomphotheriidae), *Mammut americanus* (Mammutidae), *Mammuthus columbi, M. primigenius* and *M. exilis* (Elephantidae) survived to the end of the Pleistocene (Fisher, 1996) and would have been contemporary with the human colonization of North America.

Perissodactyls (Order Perissodactyla)

Horses (Equidae)

The earliest Equidae of the early Eocene were Holarctic in distribution. The evolutionary trend across the Miocene and into the Pliocene was from woodland, through savannah to a predominance of grassland species (MacFadden, 1992). Two genera are of concern to us, having been potentially in contact with humans in the Quaternary: *Hipparion* and *Equus*. The *Hipparions* were three-toed grazers that first appeared in the middle Miocene of North America around 15 Myr (MacFadden, 1992). The group dispersed rapidly and reached the Old World via the Bering bridge by 11.5 Myr and radiated. This mid–late Miocene expansion reached Africa. This genus reached its peak in the Pliocene and subsequently entered decline, reputedly due to competition from *Equus* (Anderson, 1984). In North America the genus went extinct in the middle Pliocene and the last Eurasian species reached the lower Villafranchian (Guérin & Patou-Mathis, 1996). Potential contact with humans was therefore only possible in Africa where the genus, adapted to living in herds in the savannahs (Anderson, 1984), persisted into the late Pleistocene with eight known Plio-Pleistocene species (Bromage & Schrenk, 1999). The extinction in Africa of the most advanced and widespread species *H. libycum* is attributed to the radiation of the antelopes in the savannahs (Anderson, 1984).

The single-toed equine horses also radiated into North and Central America but did not disperse into the Old World until the late Pliocene, when the genus *Equus* (that had emerged in North America in the middle Pliocene – *E. simplicidens*) colonised Eurasia (*E. stenonis*) and Africa (MacFadden, 1992). *Equus* radiated into North America, Eurasia and Africa during the Plio-Pleistocene, the number of species depending on the authority, as species adapted to plains, savannahs and certain mountain areas (Anderson, 1984). In North America and Europe the genus became extinct at the end of the Pleistocene or just into the Holocene (e.g. *E. hydruntinus*), persisting to today in Africa and Asia. Humans would therefore have been in contact with *Equus* horses throughout the Pleistocene.

Tapirs (Tapiridae)

Tapirs are semi-aquatic mammals of moist forests. It is unlikely that these animals came into regular contact with humans on account of their geographical distribution and habitat preferences. The climatic cooling and opening up of habitats in the Pliocene and Pleistocene undoubtedly played a major role in the range contraction of tapirs that were, for example, diverse in the European Neogene. Only one species, *Tapirus arvernensis*, survived in Europe into the middle Pleistocene (Guérin & Patou-Mathis, 1996). In North America, tapirs persisted into the late Pleistocene, mostly in Florida (Anderson, 1984). Tapirs did not occur in the Plio-Pleistocene of Africa. *Tapirus indicus*, known from the middle Pleistocene (Tougard, 2001), survives today in South-east Asia and there are three other species in Central and South America.

Rhinoceroses (Rhinocerotidae)

The oldest rhinoceroses have been found in the middle Eocene of North America (around 45 Myr) and the late Eocene of Asia (around 35 Myr); Africa was not reached until well into the early Miocene (around 20 Myr) (Cerdeño, 1998). The family became extinct in North America during the Pliocene and is therefore not relevant to the present discussion. Rhinoceroses disappeared from Europe and large areas of Asia in the late Pleistocene, leaving Africa and South-east Asia with species to the present. The genus *Stephanorhinus* predominated in Pleistocene Europe, with *S. etruscus* in the Pliocene and early Pleistocene, *S. hundsheimensis* in the middle Pleistocene and *S. kirchbergensis* and *S. hemitoechus* in the middle and late Pleistocene. There were also at least three European endemic forms in the Pliocene that did not persist into the Pleistocene (Cerdeño, 1998). *Stephanorhinus kirchbergensis* was a browser of open forest and *S. hemitoechus* a grazer of open woodland and prairie (Guérin & Patou-Mathis, 1996). The woolly rhinoceros *Coelodonta antiquitatis* arrived in Europe from Asia at the end of the middle Pleistocene and also persisted until the end of the Pleistocene (Cerdeño, 1998). This was a typical cold steppe grazer whose geographical range extended across Eurasia, from Korea to Scotland and Spain (Guérin & Pathou-Mathis, 1996). Finally, the large *Elasmotherium sibiricum/caucasicum* was present in eastern Europe during the middle and early late Pleistocene and occasionally reached western Europe (Guérin & Patou-Mathis, 1996).

There was a greater diversity of taxa in Asia than in Europe during the Plio-Pleistocene: the related genera *Rhinoceros* and *Punjabitherium* from the early Pliocene; *Dicerorhinus* from the early Pleistocene; *Chilotherium* during the Pliocene; as well as the genera described for Europe (Cerdeño, 1998). *Rhinoceros unicornis* is present in the Pliocene and *R. sondaicus* from the early Pleistocene. Both survive today, in India and in South-east Asia respectively. Two species of *Dicerorhinus* are present in the early Pleistocene: *D. sumatrensis* and *D. lantianensis*. The former still survives today in South-east Asia. *Stephanorhinus etruscus* is present in Pliocene Asia and *S. hundsheimensis*, *S. hemitoechus* and *S. kirchbergensis* occur in the middle Pleistocene. *Stephanorhinus hemitoechus*, unlike in Europe, does not persist into the late Pleistocene. The other two do but not to the Pleistocene–Holocene boundary. *Chilotherium antiquitatis* appears in Asia in the middle Pleistocene, earlier than in Europe, and is preceded by the congener *C. nihowanensis* in the early Pleistocene. *Elasmotherium* persisted from the Pliocene to the late Pleistocene, *E. caucasicum* and *E. inexpectatum* in the late Pliocene/early Pleistocene and *E. sibiricum* and *E. peii* from the middle Pleistocene (Cerdeño, 1998).

The two extant African species *Diceros bicornis* and *Ceratotherium simum* were present in the Pliocene and have persisted to the present. Three other species of *Ceratotherium* were present in Pliocene Africa and another in the early Pleistocene. In North Africa *S. hemitoechus* was present in the late Pleistocene (Cerdeño, 1998). Humans would have therefore been in contact with rhinoceroses in Africa from the earliest times and in Eurasia throughout. There would never have been contact in North America.

Artiodactyls (Order Artiodactyla)

Pigs (Suidae)

The Suidae first appear in the late Eocene in Eurasia and in the Miocene of Africa. The peccaries (Tayassuidae) underwent a separate American radiation from the early Oligocene (Cifelli, 1981). In Africa the Suidae underwent a large radiation during the Pliocene and Pleistocene (Cooke 1968, 1978; White & Harris, 1977; Bishop, 1999). There were six African Plio-Pleistocene genera. *Koplochoerus* of the middle Pliocene to middle Pleistocene is thought to have been ancestral to the extant African bush pig *Potamochoerus porcus* and forest hog *Hylochoerus meinertzhageni* (White & Harris, 1977; Bishop, 1999). The genus *Metridiochoerus*, with four species from the middle Pliocene to the middle Pleistocene, is considered ancestral to the extant warthog *Phacocoerus aeithiopicus* (White & Harris, 1977; Bishop, 1999), a genus that first appears in the Pliocene (Anderson, 1984). The Pliocene *Notochoerus* that evolved from the late Moicene *Nyanzachoerus*, with three species, became extinct in the late Pliocene/early Pleistocene (Anderson, 1984; Bishop, 1999)

The Eurasian genus *Sus* first appears in the early Pliocene (Anderson, 1984) and is represented today by the widespread middle Pleistocene wild boar *S. scrofa*, and South-east Asian bearded pig *S. barbatus* (and a number of South-east Asian island forms) and also the Indian *S. salvanius*. In Europe the Villafranchian species *S. strozzii* persists locally to the start of the middle Pleistocene (Guérin & Patou-Mathis, 1996). In South-east Asia there are three other middle Pleistocene forms, *S. lydekkeri*, *S. macrognathus* and *S. officinalis* (Tougard, 2001). Frozen ground and deep snow limited the distribution of the Suidae, *Sus* being commonest in Eurasia during interglacials and interstadials (Anderson, 1984) and they never penetrated the high latitude 'Bering filter' (Flynn *et al.*, 1991) to reach North America in the Pleistocene. There is no evidence of human association with the Pleistocene peccaries (genera *Mylohyus* and *Platygonus*) (Anderson, 1984) but there must have been some contact with the recent forms of *Tayassu*. Pigs would have therefore been a regular feature of the mid- and low-latitude open woodland and shrubland landscapes occupied by humans throughout the Pleistocene in Africa and Eurasia.

Hippopotamuses (Hippopotamidae)

The family originates in the middle Miocene of Africa with the genus *Kenyapotamus* (Pickford & Morales, 1994). The genus *Hexaprotodon*, that includes the extant pygmy hippopotamus *H. liberiensis* (also placed under the separate genus *Choeropsis*), is of late Miocene origin while *Hippopotamus* is a Pliocene genus (Pickford & Morales, 1994). *Hexaprotodon* was a dominant genus in the Tertiary with a distribution that included Europe, South and South-east Asia as well as Africa. In Asia the genus persisted into the middle Pleistocene (Anderson, 1984; Guérin & Patou-Matthis, 1996; Tougard, 2001). Their African decline – most were extinct by the Plio-Pleistocene boundary – is related to the rise of the bovids that outcompeted them for food and habitat (Anderson, 1984). On the other hand, the amphibious *Hippopotamus* did not compete with the bovids. Four species are identified from Plio-Pleistocene Africa, *H. gorgops, H. aethiopicus, H. protamphibius* and *H. amphibius*, which has survived to today (Bromage & Schrenk, 1999). In India, *Hippopotamus* is present in the early Pleistocene (Anderson, 1984). In Europe, hippopotamuses had been absent throughout the Villafranchian but re-appeared in the middle Pleistocene and persisted until the start of the last glaciation (Guérin & Patou-Mathis, 1996). *Hippopotamus major* (=*antiquus*), considered by some authors to be a large form of *H. amphibius* (Anderson, 1984), is a middle Pleistocene form. *Hippopotamus incognitos* is regarded as a separate middle Pleistocene species that persisted to the last glaciation (Guérin & Patou-Mathis, 1996). *Hippopotamus* is an interglacial species outside Africa. Humans would therefore have been in potential contact with hippopotamuses throughout the Pleistocene in Africa, for much of it in South-east Asia and elsewhere in Eurasia during intarglacials. As with the Suidae, hippopotamuses never reached North America.

Camels (Camelidae)

Contact between humans and camels is unlikely to have been major during the Pleistocene, and only when humans penetrated open desert and steppic environments. Camels first appeared in the Eocene (40–45 Myr BP) of North America (Stanley *et al.*, 1994) and experienced a major radiation in the Pliocene and Pleistocene (Cifelli, 1981). Migration into Asia occurred at 3 Myr BP (Stanley *et al.*, 1994). *Camelus*, represented by the extant dromedary *C. dromedarius* that is no longer found in the wild and by the bactrian camel *C. bactrianus* that may survive in Central Asia, was present in Pleistocene North Africa and Asia (Anderson, 1984).

In North America, the camel *Camelops* and the llamas *Hemiauchenia* and *Palaeolama* persisted until the end of the Pleistocene but their remains are rare in Palaeo-Indian sites (Anderson, 1984). The giant camel genus *Titanotylopus* became extinct in the Middle Pleistocene.

Chevrotain and mouse deer (Tragulidae)

The Chevrotains first appear in the fossil record in the middle Miocene (Cifelli, 1981) and they are the first group to diverge among the ruminants (Nikaido *et al.*, 1999). There is very little fossil information available. The two extant South-east Asian species, the greater mouse deer *Tragulus napu* and the lesser mouse deer *T.* javanicus, are present in late Pleistocene sites (Tougard, 2001). Currently the family has a single representative in Africa and three in South and South-east Asia. They are forest species and are unlikely to have been frequently met by humans on account of habitat differences.

Giraffes and okapi (Giraffidae)

The Giraffidae first appear in the Miocene of Africa and radiate during the Pliocene with a significant decline in diversity in the Pleistocene (Cifelli, 1981). The genus *Giraffa* reached southern Europe and Asia in the Pliocene, remaining in the latter continent into the late Pleistocene and becoming extinct in North Africa in the Holocene (Anderson, 1984). Five species have been described from the Plio-Pleistocene of Africa (Bromage & Schrenk, 1999) although their status as variants of the extant *G. camelopardalis* is unclear (Anderson, 1984). The okapi *Okapia jonhstoni* is a rainforest species from the early Pleistocene (Anderson, 1984). It shared a common ancestor with the Pliocene *O. stillei* (Hamilton, 1978). The gigantic giraffids *Sivatherium* were present in the Pliocene of Eurasia and Africa, *S. giganteum/maurusium*, an open woodland species, persisted to the middle Pleistocene and has been associated with African Acheulian sites (Hamilton, 1978; Anderson, 1984). Giraffids, except the rainforest okapis, would therefore have been a regular feature of the open landscapes inhabited by humans in Africa, Asia and, at times, southern Europe.

Deer (Cervidae)

Deer originated in the mid-Tertiary of the Old World tropics and evolved as global temperatures and climatic stability declined, culminating with the Pleistocene glaciations. Already in the Tertiary, deer had evolved strategies that permitted them to thrive in cold climates by taking advantage of seasonal high-quality plant foods. Modern genera appeared in the Early Miocene grasslands (Geist, 1998). Three sub-families are recognised, the Cervinae (Old World deer), the Muntiacinae (muntjacs and tufted deer), and Odocoileinae (New World deer) (Miyamoto *et al.*, 1990). Some authors add the antler-less musk deer of China as a fourth subfamily – the Moschinae (Eisenberg, 1981).

The Old World deer originated in mid-Tertiary Eurasia, splitting from the muntiacines around 6–8 Myr (Miyamoto *et al.*, 1990), and subsequently spreading to Africa and India (Geist, 1998). The group saw repeated tropics to cold climate radiations during the Pleistocene. The following account is largely based on Geist (1998). The forest muntiacines appear in the late Miocene of China and are today represented by the genera *Muntiacus* and *Elaphodus* of India, China and South-east Asia. It is unlikely that these small forest deer were in any way significant as prey to hominids as other deer were, although it should be noted that they may have been regionally or locally exploited, as evidenced by their presence in South-east Asian middle and late Pleistocene sites – *M. muntjak, M. szechuanensis, Elaphodus* sp. (Tougard, 2001).

As in the case of human evolution, deer evolution proceeded from forest to open steppe types via an intermediate, ecotone, stage. This stage is represented by the sub-tropical, three-pronged, deer represented today by the genera *Axis*, *Rusa* and *Rucervus*. Fossil three-pronged deer that appear to have originated in warm climates spread widely across Eurasia in the Villafranchian, represented by *Cervus etuerarium, C. paradinensis, C.philisi, C. perolensis, Nipponicervus, Rusa* and *Rucervus*. These deer died out at the end of the Villafranchian, or the early Pleistocene in the case of the Middle East. In Japan, *Nipponicervus* survived to the end of the Pleistocene. In general terms, deer were a small element of the Villafranchian faunas. A number of advanced cervine deer also arrived in the open landscapes of Europe (some reaching China) during the Villafranchian (*Creuzetoceros ramosus, Arvenoceros ardei* and *Eucladoceros* spp.) but these did not survive into the middle Pleistocene when the cold climate megacerines and elaphines, the moose *Alces* and the roe deer *Capreolus* spread. Linked to climate deterioration was a westward intrusion of east Asian species, *C. punjabensis* reaching India around 2.5 Myr, that coincided with a major extinction of the endemic Villafranchian fauna.

Geist (1998) identifies four major Old World deer radiations that penetrated temperate and cold areas:

- The first is represented by the white-lipped deer *Przewalskium albirostris*, a high alpine steppe species of eastern Tibet.
- The second is represented by Pere David's deer *Elaphurus davidianus*, of the Chinese swamps that first appeared in the Pliocene.
- The third was the radiation of the highly successful megacerines, of which only the fallow deers (*Dama dama* and *D. mesopotamica* – considered either species or subspecies by different authors) and the Irish elk *Megaceros giganteus hibernae* survived into the Holocene. The history of the megacerines can be traced back 1.4 Myr in the eastern Mediterranean with the first reaching Europe around 700 kyr. They were also present in the Villafranchian of Japan. Megacerines were widely distributed, in North Africa, Europe, central Asia and Japan but did not reach eastern Siberia and Beringia even though *M. giganteus* inhabited the mammoth steppe. Three lines within the group entered Europe. These were: (a) the fallow deer *Dama*, the first to reach Europe in the mid-Pleistocene being *Dama (d.) clactoniana*; (b) the forest sub-genus *Megaceroides*, represented by *M. verticornis* and *M. solilhacus* in the early Pleistocene, which disappeared early in Europe but persisted (*M. algericus*) to the end of the Pleistocene in North Africa; and (c) the grassland and steppe sub-genus *Megaloceros*, with the middle Pleistocene *M. savini* and *M. antecedens* and the late Pleistocene *M. giganteus*.
- The fourth, and the most successful, was that of the genus *Cervus*. The group was represented by opportunistic, savannah, deer with some species adapting to open landscapes. Africa was not colonised except north of the Sahara. The oldest red deer *C. elephus* remains are 1 Myr and from Alaska. It is thought that the red deer originated in Asia and the species first appeared in Europe in the middle Pleistocene (700 kyr), coinciding with an arrival of a range of Asian cool temperate species. The success of the red deer, with its many forms including the North American wapiti, is due to its ecological plasticity and mixed feeding strategy. Sika deer *C. nippon* first appeared in China in the late Pliocene.

The New World deer originated in the cool temperate and cold zones of North America in the Pliocene. Miyamoto *et al.* (1990) put the split from the Old World deer lineage between 12.4 and 9.3 Myr. There are 10 extant genera: five South American and five North American, with elements in Eurasia. The latter, of concern to us here, are *Alces, Rangifer, Capreolus, Hydropotes* and *Odocoeilus*. The moose (*Alces*) branch splits off from all other New World deer in the late Tertiary. In Europe it appeared as *A. (Libralces) gallicus* in the late Pliocene around 2.0–1.6 Myr. This was the only deer to survive the Villafranchian extinctions. The ability of moose as a coloniser with rapid dispersal, related to its resource strategies that exploit instabilities in its environment, undoubtedly made it a successful lineage. *Alces gallicus* was followed in Europe by the large *A. latifrons* around 700 kyr BP, coinciding with the arrival of other Asian species including the red, roe and fallow deers and the reindeer. This species was already present in its probable centre of evolution in eastern Siberia/Beringia between 1.2 and 0.5 Myr. It spread widely across Eurasia and North America between 45° and 71° N, suggesting a cold-adapted nature although this seems difficult to reconcile with its abundance in interglacial deposits from warm areas in the middle Pleistocene. There is a similar pattern of high incidence of *A. alces* in interglacial, and absence in glacial, deposits. Moose with modern characteristics are found in deposits at the beginning of the last cold stage around 100 kyr but it is not clear if these are derived from *A. latifrons*. A similar pattern of expansion is observed in the other two North American genera that penetrated Eurasia. The reindeer *Rangifer tarandus* spread to occupy the widest longitudinal (circumpolar) distribution of any ungulate, from 14° W to 5° E. During cold episodes reindeer reached southern France and even penetrated Iberia, and were commonly hunted by humans. The roe deer *Capreolus* is the third North American cold-adapted deer. Its first appearance in Europe, as we have seen, was in the middle Pleistocene and warm climates were colonised successfully later–for example, France, Iberia and Italy around 200 kyr. Roe deer never penetrated into North Africa. Two species are recognised, the European *C. capreolus* and the Siberian *C. pygargus*. In North America *Odocoileus brachydontus* first appeared around 3.9–3.5 Myr and spread southwards reaching South America. Together with its later form, the current white-tailed deer *O. virginianus*, this lineage represents a unique success that is attributed to its opportunistic nature. The mule deer is a post-Pleistocene sub-species of the black-tailed deer *O. hemionus* thought to have hybridised with white-tailed deer.

It is clear from this brief account that deer have been ubiquitous components of the ecosystems of much of Eurasia and North America, especially since the middle Pleistocene. We will see later how particular species, especially red deer and reindeer, were a major element of the staple diet of humans in these continents throughout the Pleistocene.

Pronghorn (Antilocapridae)

The pronghorn *Antilocapra americana* of the late Pleistocene was one of the two dominant ungulates of the western North American high plains (Eisenberg, 1981). It is the only surviving species in the family. Four species in the genus *Capromeryx* ranged from California to Florida and south to Mexico from the late Pliocene to the late Pleistocene. Two species of *Stockoceros* occurred in Mexico, Arizona and New Mexico in the late Pleistocene. Five species of *Tetrameryx* occurred in Texas, California and New Mexico in the Pleistocene. *Stockoceros* remains have been found in association with humans dated at 11.5 kyr BP (Anderson, 1984).

Wild cattle and antelopes (Bovidae)

The bovid family radiated extensively in the Pliocene and, particularly, the Pleistocene (Cifelli, 1981). It is the most diverse family and naturally colonised all continents except South America and Australia. It is also the family that, along with the deer, has been central as a food resource to the hominid populations of the Pleistocene. The Bovidae is divided into nine sub-families:

(1) Alcelaphinae: wildebeests, hartebeest; Africa; 7 extant species.
(2) Antelopinae: antelopes, saiga; Africa, Asia; 38 extant species.
(3) Bovinae: kudu, eland, four-horned antelope, bison, wild cattle; Africa, Eurasia, North America; 24 extant species.
(4) Caprinae: chamois, mountain goat, serow, goral, musk ox, takin, bharal, goats, sheep; Eurasia, North America, Africa; 32 extant species.
(5) Cephalophinae: duikers; Africa; 19 extant species.
(6) Hippotraginae: sable, roan antelopes, blue buck; Africa, Asia; 7 extant species.
(7) Peleinae: rhebok; Africa; 1 extant species.
(8) Aepycerotinae: impala; Africa; 1 extant species.
(9) Reduncinae: kob, reedbuck; Africa; 8 extant species.

Representatives of all the sub-families have been found in association with hominid levels (Anderson, 1984; Bromage & Schrenk, 1999; Tougard, 2001) suggesting a long and continued interrelationship between hominids and the Bovidae.

Herbivore distribution patterns

A cluster analysis (Figure 2.2) of the Pleistocene distribution of mammalian herbivores at the level of sub-family separates geographical regions in the following manner (in order of separation):

(1) Africa.
(2) Desert regions.
(3) India, the eastern mid-latitude belt, South-east Asia and China (South-east Asia, then China, separate from the cluster).

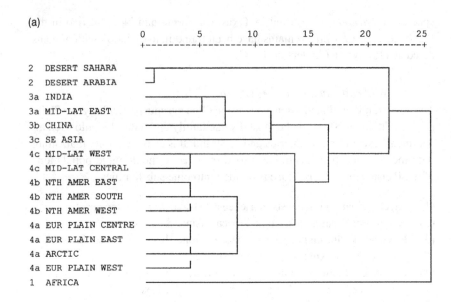

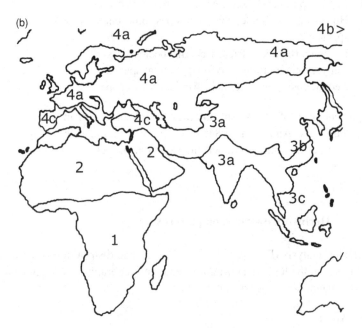

Figure 2.2. Separation of geographical regions by similarity of mammalian herbivore components. (a) Results of cluster analysis; (b) representation of results in map form.

(4) The Holarctic. The Holarctic cluster then separates out, in order, into,
   (a) the western and central mid-latitude belt;
   (b) North America;
   (c) the Eurasian Plains and the Arctic.

It is interesting to note that the separation of the western and central mid-latitude belt from its cluster occurs earlier than the separation of South-east Asia from its group, indicating its distinctiveness, and that North America separates from its cluster before China from its cluster.

There are two underlying patterns to this distribution. On the one hand there is the historical pattern of the isolation of the various regions. Thus Africa, the South-east Asian cluster and North America are distinct because of the periods of isolation caused by deserts, ice and sea barriers. South America and Australia, had they been included in the analysis, would have separated out even earlier for this reason. In this respect North America is the least distinctive and Africa the most. At another level there is the ecological pattern. The desert regions, independently of geographical location, separate out early on account of the impoverished faunas. The separation of the Eurasian Plains–Arctic–North America group from the western and central mid-latitude belt is a combination of the two patterns. It shows that the Eurasian Plains, in spite of a huge longitudinal spread, are an ecological unit that cluster with the Arctic (a northward extension of the plains) and North America (an eastern continuation). This unit is sub-divisible into Eurasian Plains/Arctic–North America for reasons of occasional isolation. The separation of the eastern mid-latitude belt from the rest of the belt, and its affinity particularly with India, reflects a shared history related to proximity. The separation of the mid-latitude belt from the Eurasian Plains is indicative of a marked ecological boundary that, as I shall develop in this book, has been crucial in human evolution.

Hominids would have encountered a group of similar mammalian herbivore size classes and ecological equivalents across their geographical range in the Pleistocene (Table 2.1 & 2.2; Figures 2.3 & 2.4). I will discuss the exploitation of prey by hominids in Chapter 5. At this stage, I will highlight some macroecological patterns. When we consider the range of open habitats (usually with few, scattered, trees and shrubs) and shrublands that hominids would have encountered in different parts of the geographical range we find recurrent trends. For huge areas of Africa, India, China, in patches within the mid-latitude belt, across the Eurasian Plains, in the Arctic and over much of North America similar mammals would have been encountered. Among the largest forms (i.e. >1000 kg) would have been the elephants (including mammoths) and the rhinoceroses. In the next size category (500–1000 kg) would have been the large bovids such as bison *Bison* spp., wild cattle *Bos* spp. or African buffalo *Syncerus* spp. The

Table 2.2. *Frequency of occurrence of mammalian herbivore taxa in African hominid and non-hominid sites (after Bromage & Schrenk, 1999). This table, together with Figure 2.4, suggests that the main taxa associated with hominids were pigs (Suidae), bovids (Bovinae), wildebeests and related forms (Alcelaphinae), giraffes (Giraffidae), horses (Equidae), and hippopotamuses (Hippopotamidae). The majority are medium-sized taxa (see text). Hippopotamuses may be associated with humans because many sites are near wetlands, where hippopotamuses preferentially would have been found, and need not indicate a hunting preference. The relationship with giraffes is unclear. A positive but not a statistically significant relationship is indicated by parentheses*

Taxon	More than expected in hominid sites	Fewer than expected in hominid sites	χ^2	Significance
Suidae	+		13.53	<0.005
Bovinae	+		8.62	<0.05
Antelopinae	(+)		5.58	NS
Reduncinae	(+)		1.31	NS
Alcelaphinae	+		10.66	<0.005
Giraffidae	+		18.5	<0.025
Equidae	+		8.61	<0.05
Elephantidae	(+)		6.94	NS
Rhinocerotidae	(+)		2.54	NS
Hippopotamidae	+		9.93	<0.025
Hippotraginae	(+)		7.81	NS
Aepycerotinae	(+)		7.53	NS
Camelidae		(+)	5.57	NS

predominant prey would have been in the 100–500 kg category. In Africa this category would have been dominated by bovids such as gnus *Connochaetes* sp., hartebeests *Damalops* and *Alcelaphus* spp., or kob *Kobus* sp. In Eurasia and North America this category was largely filled by the deer *Cervus* and *Rangifer* spp. Horses *Equus* spp. would have been present throughout. The smaller category (<100 kg) would have included the smaller African bovids such as steenbok *Raphicerus* sp., springbok *Antidorcas* sp. and the gazelles *Gazella* sp. that also occupied large areas of southern Asia. In Eurasia the group would have included the fallow deer *Dama* sp. and the saiga antelope *Saiga* sp. Pigs such as boar *Sus* sp. in Eurasia, warthog *Phacocoerus* sp. in Africa or peccary *Platygonus* sp. in North America would have largely fallen into this category. These animals, occurring as similar or different species across Africa, Eurasia and North America, would have been those available to all hominid populations throughout their range.

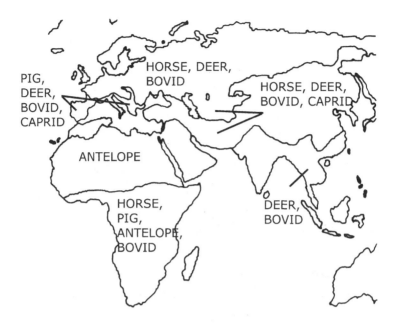

Figure 2.3. Main mammalian herbivore components by geographical region.

The large tracts of desert would have been impoverished relative to these open habitats. The main species would have been adapted to arid conditions. Some, such as gazelles, *Gazella* spp., would have been encountered in different parts of the range. Others, such as camels, *Camelus* spp., in Eurasia or *Camelops* sp., in North America, or oryx *Oryx* sp., would have been geographically localised.

Dense temperate, tropical and equatorial forests were not regularly-used hominid habitats (Gamble, 1993). Mammalian herbivores do not reach the high diversity, densities or levels of aggregation that they do for example in open grasslands (Bigalke, 1968; Jarman & Sinclair, 1979; Prins & Reitsma, 1989; Srikosamatara, 1993; Fritz & Duncan, 1994; Fryxell, 1995; Prins & Olff, 1998), and prey are more difficult to detect and catch. They are often of small size. It is not surprising, therefore, that forest herbivores, e.g. roe deer *Capreolus capreolus* in Eurasia, duikers *Cephalophus* spp. in Africa or muntjacs *Muntiacus* sp. in South-east Asia, do not feature as prominently in hominid sites as do open habitat herbivores.

Independently of region, and at smaller spatial scales, there are habitats that have been favoured by hominids on account of high prey densities or because they hold particular species. Such habitats are readily identifiable by their features in the landscape. These are:

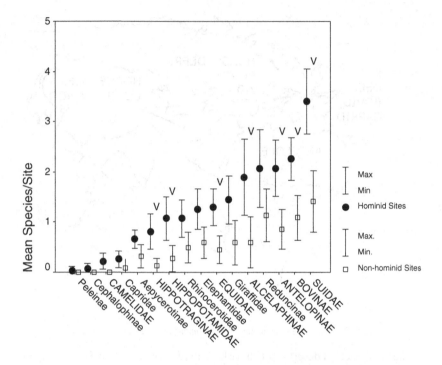

Figure 2.4. Comparison of mean (with 95% confidence limits) number of major mammalian herbivore species in African localities with hominids and without hominids (after Bromage & Schrenk, 1999). Arrowheads point to taxa in which statistically significant differences are found with more species in sites associated with hominids than in those without.

(1) Rocky habitats that usually hold small- to medium-sized mammalian herbivores. The Eurasian mid-latitude belt appears particularly favourable throughout with, for example, ibexes *Capra* spp. and chamois *Rucicapra* sp. Some *Capra* have penetrated as far as the north-eastern coast of Africa. The klipspringer *Oreotragus* sp. in Africa and the mountain goat *Oreamnos* sp. in North America, occupy similar habitat.

(2) Wetlands would have acted as focal, usually seasonal, points of concentration of mammalian herbivores as well as other animals (e.g. wildfowl, reptiles, amphibians, fish). The species range present in each site would be dependent on geographical location and surrounding habitat. In general terms, the species typical of open landscapes would be expected. Hippopotamus *Hippopotamus* sp. would have been a mammalian herbivore typically associated with such habitats in warm areas.

(3) Coastal habitats would have attracted a particular group of animals. It is becoming increasingly apparent that hominids regularly exploited such habitats in different parts of the world (Klein, 1999; Stiner, 1994, Finlayson *et al.*, 2000b). Marine mammals may have been largely acquired by scavenging but there are indications that suggest that seals may have been selectively harvested at particular times of the year (Balter, 2001). Harvesting marine mammals by hominids may have been localised or may have been a more generalised habit. In addition to marine mammals, molluscs were harvested in a number of coastal regions, for example in the Mediterranean and Atlantic (Stiner, 1994; Finlayson *et al.*, 2000b).

Mammalian herbivore biogeographical patterns and climate

Climate change would have been largely responsible, through habitat change, for modifications in the distribution patterns of mammalian herbivores. These would have largely taken the form of expansions, contractions or shifts in the range of each species in response to their particular requirements (FAUNMAP, 1996). The similarity of requirements of a number of species, or their interdependence, would produce patterns of apparent movements of faunas as is often reported in the literature (Tchernov 1992, 1998) but this would be no more than coincidence in response to the same ecological parameters. At other times, species' responses would have varied and the effects on geographical distribution would also have been different, thus producing apparently unusual combinations of species in particular areas. The intensity and range of such responses would have in addition varied depending on the particular history of each population. Thus, two populations of the same species might show a different level of response to the same signal at different times because of such variations as starting population size or geographical range centre. It is for these reasons that animal communities can be so different through time and why past communities may have no present-day analogues (Guthrie, 1984). The possible combinations are virtually infinite.

Despite these difficulties it is possible to establish certain patterns. At the higher taxonomic levels (such as family and sub-family) the distribution of, for example, mammalian herbivores is fairly constant in spite of large climatic and environmental fluctuations (Table 2.1; Figure 2.3). There are, for example, elephants across much of Eurasia, Africa and North America throughout, so hominids would encounter this herbivore type in suitable habitats across much of their range. The same would apply to the other herbivore resource types discussed earlier in this chapter – i.e. rhinoceros, large bovid, horse, medium bovid/deer, small bovid/deer and pig. The exceptions would have been the ice, sandy and stony deserts. As these deserts retreated under particular climatic circumstances (Chapter 6), higher order herbivore types that had been absent would have had the opportunity to enter. This would have occurred, for example, when tropical African mammals occupied areas of the Sahara or when Eurasian Plains herbivores colonised boreal regions.

The responses (range changes or extinctions) to climatic and environmental fluctuations would therefore have been more frequent at lower taxonomic levels and detectable at the species and, less frequently, the generic levels. The responses would be even more frequent, but more difficult to detect, in local populations of a species. In the case of mammalian herbivores such changes could also have had an impact on predators, including hominids, in cases where different species had different behaviours. In terms of the strategies required to exploit them, the migratory reindeer of the plains were a very different resource from the red deer of the woodlands. So, for example, although medium-sized deer would have been available to hominids in southern France for much of the Middle and Upper Pleistocene, the actual species present would have varied considerably depending on climate-induced habitat changes (Mellars, 1996). I will discuss the impacts of such changes on hominids at the level of species and populations in Chapter 5.

Main trends in mammalian herbivore range shifts that would have been significant to hominids at the macro-ecological level would have been:

(1)   The north–south range changes of tropical savannah and grassland African species in relation to the expansion and contraction of the Sahara.

(2)   The north–south range changes in Arctic species in relation to the expansion and contraction of the glacial ice sheets and the tundra.

(3)   The east–west range changes of steppe-adapted species across the Eurasian Plain in relation to the expansion and contraction of the steppe.

(4)   The range contractions and expansions of tropical savannah and grassland species in Africa, South and South-east Asia in relation to the expansion and contraction of the rainforest.

(5)   The north–south range contractions and expansions of temperate species into and out of mid-latitude refugia in relation to changes in the distribution of forest and shrublands.

(6)   At a smaller spatial scale, the altitudinal shifts in the distribution of species in topographically heterogeneous landscapes in response to temperature controlled vegetation changes.

Synthesis

We may therefore identify the following trends that will assist us in our synthesis of human evolution in the following chapters.

Mammalian herbivores as a resource

The medium-sized (100–1000 kg) packages of protein that were consumed by humans during the Pleistocene were available world-wide and throughout the Pleistocene. Minor differences in taxa separated the regions. The constancy in availability of these packages permitted humans to adapt for their exploitation. The robusticity of many human populations reflects such adaptation to hunting these mammalian herbivores. The loss of mammalian herbivore diversity and biomass at the end of the Pleistocene led to the herding and domestication of certain species, in particular the goat, sheep, cattle, pig and horse (Diamond, 1997). This may be considered as a latter form of adapting to this resource.

Topography of the terrain as a template in which to obtain mammalian herbivores

The topography of the terrain at large scales has been another constant through-out the Quaternary. Humans, depending on where they settled, were therefore able to adapt to the types of terrain in which they were able to make a living. Those in Africa adapted to a mix of flat and broken landscapes. As they expanded away from Africa they would have adapted further to the specificities of the terrain colonised. Those reaching the mid-latitude belt increasingly adapted to broken terrain while those, much later, reaching the plains of Eurasia would have increasingly adapted to flat landscapes. In either case, and also in terms of ability to obtain mammalian herbivores, existing African templates would have pre-adapted humans to ex-African colonisations.

Habitats and landscapes

The distribution of landscapes and habitats changed frequently during the Quaternary in response to climate change. The rate of change intensified through the Pleistocene. Humans were initially adapted to intermediate habitats (those between closed forest and open plains) and mosaic landscapes. Through time they increasingly adapted to open habitats and plains landscapes. Adaptation to intermediate and then open habitats and landscapes enabled range expansions in humans. Because habitats and landscapes frequently changed spatial location due to climate changes, humans instead of adapting *in situ* tracked these changes. This tracking across spatio-temporal scales marked the range expansions, shifts and contractions that I will describe in Chapter 3.

Barriers

Desert and sea, in particular, acted as physical and ecological barriers to range expansion or shift. Their extent, from total impermeability to high porosity, varied with climate change. Cold and aridity intensified desert barriers but reduced marine barriers. Some, like the deep Strait of Gibraltar, were barriers throughout. Barriers can also occur at smaller spatial scales, as between habitat patches, the scale of an individual's perception of the environment being critical in the ability to overcome such barriers (Wiens, 1997).

Tropical–boreal gradients characterised human evolution and acted as barriers that were progressively dismantled. Throughout the Quaternary, tropical environments have provided greater resource diversity than boreal ones. Human resource consumption is therefore expected always to present a profile of decreasing diversity from the tropics towards higher latitudes. Temperate environments have occupied an intermediate position. Humans have adapted in the form of an omnivore–carnivore gradient. Colonisation of boreal environments, in particular, required almost total carnivory. There is a corresponding daylength seasonality regime between the tropics and the polar regions. These constants have been critical in the adaptation of humans to non-tropical environments.

In the following chapters, these trends will allow us to interpret human evolution in the Quaternary. Humans adapted to obtaining mammalian herbivores in rugged terrain in intermediate habitats and landscapes. In time they also adapted to resource-rich environments characterised by flat terrain and open landscapes. The changes in habitat and landscape modulated the spatial distribution of populations that were driven by these adaptations to resource and terrain. Barriers temporarily suspended this dynamic system.

3 Human range expansions, contractions and extinctions

African beginnings

We saw in the previous chapter how climate-induced habitat and landscape changes acted as catalysts to human range expansions, shifts and contractions. This chapter explores these processes in greater detail.

Novelty in the genus *Homo* was generated repeatedly in eastern Africa. It is not unexpected that the sources of biological novelty should be tropical given the tropical nature of the primates as a whole (Foley, 1987) and the scarcity of species that reach away from the tropics. The nature of the distribution of the ape lineage and the distribution of open savannah-type habitats close to tropical forest make an African tropical origin a virtual certainty and this is well supported by the fossil evidence (Stringer & Gamble, 1993; Akazawa, 1996a; Klein, 1999). Novelties arose repeatedly in the *Homo* lineage but they also occurred deeper in time with the adaptations to open environments and departures from frugivory and herbivory of the pre-*Homo* forms.

Arid communities became established in East Africa by 23 Myr and there has been no real change in vegetation in the past 15.5 Myr. C4 vegetation, characterised by grasses and sedges in warm arid, open, habitats, appeared around 15.3 Myr BP (Kingston *et al.*, 1994). Early hominid evolution took place in sub-Saharan Africa in situations of increasing environmental instability that led to forest contraction and at the expense of open environments (Foley, 1987; Foley & Lee, 1989; Bobe *et al.*, 2002). Early hominid evolution took place within the context of a heterogeneous mosaic of environments in intermediate situations between the closed forest and the open grasslands (Kingston *et al.*, 1994). Tropical climate was cool and variable during glacial cycles (Schrag *et al.*, 1996; Thompson *et al.*, 1997; Webb *et al.*, 1997; Bush & Philander, 1998). These effects, generated by the expansion of the high latitude ice sheets, were met, especially in the last 1 Myr, by abrupt vegetation distribution changes largely caused by increased aridity (deMenocal, 1995; Hughen *et al.*, 1996). These changes led to the progressive expansion of open vegetation that became central to human evolution. Occupation of open habitats required abilities for coping with increasing uncertainty, seasonality and patchiness. These needs were met by morphological and behavioural responses (Foley, 1992;

deMenocal, 1995; Stanford, 1999; Stanford & Bunn, 2001). The advantage of these open environments centred around their high net primary productivity (Field *et al.*, 1998) and the relative ease of locating aggregations of mammalian herbivores.

Evolution in the genus *Homo* is marked by a series of adaptations that improved performance in the arid and open terrestrial environments of Africa, and conditioned the course of that evolution (Foley, 1987, 1992; Foley & Lee, 1989). Beginning with bipedality (Jablonski & Chaplin, 1993; Hunt, 1994, Isbell & Young, 1996; Richmond *et al.*, 2001), these adaptations were morphological and behavioural, including in its most sophisticated form, culture (Wheeler, 1994, 1996; Queiroz do Amaral, 1996; McHenry & Berger, 1998). Brain enlargement, reaching a maximum in the archaic humans (e.g. the Neanderthals) and in the Modern Humans, stands out among these adaptations (Aiello & Dean, 1990; McHenry, 1994; Aiello & Wheeler, 1995; Kappelman, 1996; Elton *et al.*, 2001). We can attribute these adaptations to increasing specialisation to the open and arid environments of tropical Africa (Reed, 1997) and they improved resource acquisition and, incidentally, permitted expansion outside Africa. Such features included increased body size and a long-limbed morphology that enabled changes in ecology, including larger home ranges and increased dispersal ability (Anton *et al.*, 2002) and the capacity for endurance running (Carrier, 1984). They pre-adapted populations for success in other open environments, reaching the maximal expression in the late Pleistocene on the Eurasian Plain (Gamble, 1986, 1993, 1999).

The dynamics of colonisation and extinction

Straus & Bar-Yosef (2001) have defined hominins as 'purpose-driven species'. The notion of purpose-driven human migrations is pervasive in the literature. For many authors, the migrations into Australia must have required craft and navigation skills (Klein, 1999). These same authors seem to ignore the ability of other primates (e.g. macaques) that, without boats, regularly colonised deep water islands in South-east Asia that were never linked to the mainland (Brandon-Jones, 1996; Abegg & Thierry, 2002). Dispersals out of Africa are also referred to as migrations. This way of defining changes in the geographical range of humans through time confuses proximate factors, such as curiosity, with the ultimate factors responsible for range changes. Behind the arguments is the notion that humans are apart from all other living organisms and that 'special' mechanisms can be found to explain their behaviour. If this had indeed been the case then we would have to postulate a non-biological model of human geographical expansion, one that would have been independent of natural selection. Given that humans behaved as components of the ecosystems of which they were a part, it is far more likely that they were ultimately very much governed by selective pressures even if their socio-cultural attributes (themselves phenotypic expressions of an evolved genetic plasticity) gave them significant advantages over other species in the same ecosystems.

So how do changes in geographical range occur? They are the response to demographic pressure within the existing range and favourable environmental changes in the periphery (Foley, 1997; Dynesius & Jansson, 2000). The speed of invasion into a new area is the product of the interaction between local adaptation and genetic and demographic parameters (Kirkpatrick & Barton, 1997; Gárcia-Ramos & Rodriguez, 2002). A population may be increasing as a result of favourable conditions and intra-specific competition forces some individuals to disperse away from the dense core area. Most dispersers will not find a suitable area for settling or may end up in an area already occupied by the same species. New marginal populations may, however, occupy optimal habitats or they may occupy sub-optimal ones in which they are nevertheless able to make a living. A successful colonisation depends on the capacity to adjust genetically (Gárcia-Ramos & Rodriguez, 2002) or behaviourally to a spatially varying environment and results in a longer lasting population that is independent of arrivals of additional dispersers (Gyllenberg *et al.*, 1997). It is worth noting in this context that sink populations that are maintained by immigration from source populations, even when birth rate is below death rate, may show varying degrees of permanence (Brown & Kodric-Brown, 1977; Dytham, 2000). Thus, the presence of a population in an area is not necessarily proof of its success in that area. We should be conscious of this when considering individual archaeological sites. At the other end of the spectrum, colonisation–extinction models predict that at any point in time there will be a proportion of habitable patches that will be empty because of demographic stochastic extinctions (Hanski & Gilpin, 1997; Tilman & Karieva, 1997; Hutchings *et al.*, 2000). So absence is not proof of unsuitability either!

Returning to a successful colonisation, as the population grows so its range expands until unsuitable habitats are encountered, unless individuals are able to adapt to the new circumstances. Ranges may also shift. If conditions on one side of the range are deteriorating then the population contracts in those areas, either through local extinction or by movements of individuals into core areas with consequent increase in intra-specific competition. In such cases the advantage is likely to be with the residents and so the local marginal populations may become extinct anyway. If, at the same time, favourable conditions are becoming available (perhaps due to a climate change) then there will be expansion into those new areas by the same process described before. The outcome is a range shift. These are generalised models. There are other ways of expanding geographical range. Central areas in the range need not necessarily be the core population areas and individuals may 'jump' from one optimal habitat to another even if there is unsuitable habitat in between (Lewis, 1997). Hewitt (1999, 2000) considers that populations on the northern edge of a refugium would have rapidly recolonised empty territory during climatic amelioration, with the leading-edge expansion being led by long-distance dispersers rapidly setting up colonies and expanding. Such expansions would necessarily lead to loss of genetic diversity among these small founder populations.

Highly vagile animals are able to integrate heterogeneity over broader scales and therefore perceive the environment with a coarser filter (Wiens, 1997). Dispersal ability and dispersal rate are therefore important internal population parameters (Lehman & Tilman, 1997; Lewis, 1997). Colonists can arrive actively or passively and there may be a number of reasons why they arrive in a new area: (a) following a change of conditions; (b) following removal of a barrier; (c) following the creation of a passageway; or (d) following a genetic change which adapted them to conditions in the colonised area. In cases of environmental instability, as in Pleistocene Eurasia, the time delay of population response, relative to the period of the environmental cycle, is crucial for persistence. Populations with fast response that track cycles will reach periodic lows and risk extinction. The Neanderthals are a good example. Populations with slow response may be able to keep a more or less stable population size. The Moderns may well be an example (Chapters 5 and 7).

One way of reducing the effects of environmental fluctuations is to prolong the response time to environmental changes (Hutchings *et al.*, 2000) – i.e. to invest in environmental resistance. This can be achieved through 'escape responses'. Dormancy or hibernation are examples. I argue in this book that the complex social systems of Moderns, their extended networks and their systems of operating at large scales and storing and caching resources effectively prolonged their response to environmental changes, that is Moderns invested in environmental resistance.

Temporal and spatial heterogeneity are likely to be perceived by a colonist population as being greater than in the source area. This means that during an initial phase of colonisation a population needs to rapidly colonise many patches to reduce the risk of extinction. In spatial terms, dispersal ability can be an escape mechanism. A high instantaneous rate of increase (r), an avoidance of density-dependence and high dispersability all guarantee successful colonisation in environments expected to fluctuate either systematically, randomly or spatially. Competition can alter the success of colonisation. A species with potential to change its position along the resource spectrum is likely to be a good coloniser. We see these attributes in the characteristics of Moderns and we also have a theoretical basis for understanding their eventual success in areas like the Middle East where they may have faced competition from Neanderthals (Chapter 7).

Finlayson *et al.* (2000a) have proposed the generalised conditions that would have lead to geographical range expansions and contractions on a global level during the Quaternary. These range changes have to be viewed against the climatic backdrop that characterises the Quaternary and differentiates it from earlier periods (Denton, 1999). Throughout the Quaternary we observe cyclical climatic changes, their frequency intensifying towards the latter stages (Imbrie *et al.*, 1984; Ruddiman *et al.*, 1986). We observe, at different scales, variability even in equatorial and tropical regions (deMenocal, 1995). It is this climatic variability that, through consequent habitat variability, drove the dynamics of geographical range in humans and indeed in many other species (Potts, 1996a, b, 1998). Given that the number of such major and minor oscillations was very high over the last two million years (Shackleton & Opdyke, 1973, 1976; Shackleton *et al.*, 1984) we would predict many geographical expansion and contraction events, not just one or two. The intensity and duration of each event, coupled with the demographic situation of the initial population in the core area, would have been the key elements in the extent and direction of the range expansion (Finlayson *et al.*, 2000a). Once populations became established away from the initial core area then, assuming they survived subsequent unfavourable events, these secondary core populations would have acted as new sources of expansion when favourable conditions resumed.

This leads me to the all-important question of extinctions. As with range expansions we have to view extinctions at different scales. At the smallest scales, extinctions of local populations would have been a regular feature of human populations throughout the Quaternary. Such extinctions would have probably affected marginal populations most severely and small effective population sizes would have meant that many extinctions would have been the result of stochastic processes (see Chapter 7). Regional extinctions would have been less frequent, though not uncommon, and would have occurred when more significant alterations in favourable conditions happened, sufficient for all the local populations within a region to have been affected. Finally, global extinctions would have been the least likely given that regional populations somewhere would have been buffered against unfavourable conditions elsewhere.

Human populations in tropical and equatorial regions would have been least prone to extinction given that the range of resource options in such regions would have been greatest and the effects of climatic oscillations on habitats least felt (Figure 3.1; Finlayson *et al.*, 2000a). In addition, these areas would have enjoyed a fairly constant day length (and therefore year-round foraging and hunting) throughout the year. So populations in equatorial and tropical Africa, and subsequently in South-east Asia, would have enjoyed the greatest degree of regional permanence. Next would be the proximal warm temperate regions and the least conducive to regional permanence would have been the cool temperate and boreal regions. As humans evolved physical and behavioural

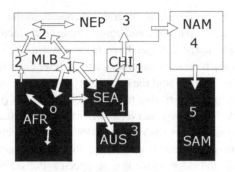

Figure 3.1. Source and sink regions in human evolution. Arrows indicate probable strength and direction of geographical expansion. The Strait of Gibraltar as an entry point is only partly supported by the available evidence (see text). The boxes represent major regions of the world. Largely tropical areas are in black: AFR, Africa; SEA, South-east Asia; AUS, Australia; SAM, South America. Temperate areas are in white: MLB, Mid-latitude belt of Eurasia; CHI, China; NEP, North Eurasian Plain; NAM, North America. Numbers indicate the approximate process of initial colonisation by *Homo*. For any given stage in the colonisation process of *Homo*, persistence is predicted to be highest in black (source) areas and lowest in white (sink) areas. Australia and South America were colonised too recently to have been important source areas in the Pleistocene. Mid-latitude Eurasia and China act as refugia and secondary sources of colonisations of areas to the north. Only Africa and South-east Asia would have had continuous occupation after 1.9 Myr BP. After Finlayson *et al.* (2000a).

adaptations that improved colonisation and persistence so areas further away from the tropics could be successfully colonised, Moderns being the best at doing so.

Viewed in this manner the extinction of the Neanderthals (Chapter 7) is not unusual or even surprising. It is the extinction of a complex of regional populations in Europe and western Asia. It is an example of events that probably occurred repeatedly earlier in the Quaternary and tells us that we must exercise care in taking for granted cases of regional continuity in non-tropical areas. Take the case of *H. antecessor* at Atapuerca (Spain) 800 kyr ago (Carbonell *et al.*, 1995). Were these the ancestors of subsequent European humans or did they simply go extinct? The serious answer to this question is that we do not know. Yet, on morphology (in spite of the inherent problems with morphological criteria, Chapter 4) a direct ancestry is proposed. But even in Atapuerca itself we cannot convincingly show continuity. The fossils from Gran Dolina and Sima de los Huesos (Spain; Arsuaga *et al.*, 1993) are separated by half-a-million years and we simply do not know what happened in between. Hopefully, with time we may know as excavations proceed but today we cannot say one way or the other with certainty. In ecological terms it is of interest to note that when humans lived in Atapuerca, climatic conditions were milder than at present (Cuenca-Bescós *et al.*, 1999; Cuenca-Bescós, 2003; van der Made, 1999). Today, Atapuerca is a harsh environment in the winter and it must have been even harsher during glacials. To suggest continuity is, to my mind, a very bold assertion in the light of the limited data available.

So if there were multiple colonisations and extinctions in Eurasia, how many were there? At present, that is an impossible question to answer. The evidence from Orce (Spain) is unclear but suggests a possible earlier colonisation that may have occurred via the Strait of Gibraltar (Arribas & Palmqvist, 1999; Oms *et al.*, 2000). That is open to debate and must await further evidence. We would then have to see if these humans were part of the same colonisation that lead to Atapuerca or something else. Post-Atapuerca there may have been several colonisations of Europe, each time with greater success. The pre-Neanderthals and the Moderns were the last two of a chain.

The colonisations would have been part of a continuum of range expansions of varying extent, local and regional extinctions, subsequent re-colonisations and even re-colonisations into areas occupied by a previous colonisation that persisted. The latter, I would predict, would have been most frequent close to the tropical core areas. In such areas of contact the outcome would have been determined by a variety of factors including the time and degree to which the two meeting populations had been previously isolated, and thus the degree of genetic, morphological and behavioural isolation, the densities of the two populations relative to environmental carrying capacity and the degree of ecological isolation. In cases where the conditions for competition would have been right, then population attributes that gave one population the edge over the other would have been critical. In the rapidly fluctuating conditions of the Quaternary, the conditions for such competition would have been rare, more so as one went away from the tropics.

The global pattern of colonisation and extinction

The patterns of faunal interchange between tropical and boreal regions have a deep history within the Neogene (Pickford & Morales, 1994). Latitudinal fluctuations in the boundary zone between the tropical and boreal biogeographical realms have marked the past 22.5 Myr. The difference in receipt of solar energy on the Earth's surface and the inclination, at a steep angle to its orbital plane, of the axis of the Earth's rotation have meant that the zone of maximum receipt of solar energy shifted latitudinally across the globe causing seasonality. Seasonality at high latitudes is overwhelmed by daylength and temperature changes (Pickford & Morales, 1994). Migration, hibernation and summer reproduction are typical responses of animals to these predictable changes. Humidity changes dominate the low latitudes where temperature and daylength variations are of lesser importance. Wet and dry seasons thus dominate tropical seasonality patterns. Aestivation and wet season reproduction are typical responses.

Throughout the Pleistocene the populations of humans across the world underwent fluctuations, range expansions and contractions. In this respect they differed little from a whole range of organisms (Hewitt, 2000). Those at greatest risk of extinction were those furthest away from the tropics, the habitat fragmentation caused by increasing cooling and aridity contracting the northern parts of the range and also compressing the altitude range. The length of such adverse climatic periods, occurring as single events or series of such events with brief interludes, was probably more significant than the intensity of the adverse pulses. Range contraction would have taken the form of regional population extinctions especially when climate variations were rapid (Hewitt, 1996, 1999, 2000), a situation that caused the extinction of, for example, tree species (McGlone, 1996), reptiles (Busack, 1986) and mammals (Martin & Klein, 1984). During improved climatic conditions, northward extensions of the range of populations that had managed to survive commenced from southern refugia (Hewitt, 1999, 2000). The risk of becoming extinct would have depended on: (a) the ability to colonise sufficient sites during periods of peak abundance so as to permit survival when they became rare; (b) stochastic effects that might have eliminated populations that spent long periods in small isolated sites; and (c) the ability to track suitable climates during periods of rapid change (McGlone, 1996). In the case of trees, for example, differences in source areas and migration rates continuously changed the forest composition north of the Alps (Zagwijn, 1992). Faunal composition would have varied similarly as animals behaved in a Gleasonian manner, that is individually responding to environmental variables (FAUNMAP, 1996; Hewitt, 1999; Chapter 2).

Tropical African hominid populations would have benefited from increased cooling and aridity and their range would have expanded within the tropics. Subsequent amelioration immediately after cold/arid periods (when populations were at their highest) would have permitted northward expansions as the Sahara Desert became savannah and grassland (Finlayson *et al.*, 2000a). In this way, tropical human populations would have repeatedly reached south-west Asia, the range expansion sometimes being checked by changing climate.

Geographical barriers would have then played a major role in the continuing expansion of the geographical range. In the west, the Strait of Gibraltar appears to have acted as a barrier on a number of occasions but not necessarily always (Finlayson *et al.*, 2000a). Thus the similarity in Acheulian technology on the two shores of the Strait has led some authors to postulate that movement did occur at such times (Alimen, 1975; Giles Pacheco & Santiago Pérez, 1987). In the east, the barriers of the Taurus, Pontic, Zagros and Caucasus would have checked the expansion of the tropical humans (Finlayson *et al.*, 2000a). Warm conditions would have restricted movement here because much of the landscape would have been densely wooded and unsuitable for expanding human populations. Cold conditions would have been just as unsuitable as montane habitats reached close to the shore. Passage could have occurred along river valleys or along the extended coastal shelf during intermediate climates, most probably immediately after cold phases when southern populations would have been augmenting and the forests had not closed up. Passage east from the Middle East or the Horn of Africa would have been much easier. The eastward spread would have kept populations south of the Himalayan mountain mass and within tropical or semi-tropical climates. Finlayson *et al.* (2000a) predicted that the frequency of range expansions from Africa into different parts of the world would have followed the sequence (Figure 3.1) discussed below.

Sahara, Middle East and southern Africa

These areas would have received expanding populations most frequently on account of similarity of climate and proximity to source areas. North-west Africa, however, belonged in the next category because of the combined effect of distance from source areas and the Sahara Desert.

South to South-east Asia, north-west Africa and south-east Europe

These areas would have been next in frequency of colonisations because of climatic similarity, and relative ease of access. Some areas would have been relatively close to source areas but others relatively distant. Australia is a natural extension of this belt on the South-east Asian side but would only have been colonised once the sea barrier could be overcome, occurring substantially earlier than 50 kyr (Thorne *et al.*, 1999; Bowler *et al.*, 2003). South-east Europe falls into this category because of climate similarity and proximity and does not appear in the previous category with the Middle East because of the effect of the intervening mountain barriers.

Central and western Mediterranean Europe and the Eurasian Plain

These would have been the next areas to be colonised, increasing distance from source areas and mountain barriers, delaying access. In the earlier colonisations only the Mediterranean and adjacent lands were colonised, the Plains being the last to be reached (Chapter 7). Climatic difference from source areas appears not to have been an impediment, at least to the later expanding populations, because of the structural similarity of these environments to those of the source areas and also the availability of mammalian herbivores (Chapter 2). The climatically more suitable areas of the central and western Mediterranean are also included in this category because of their distance from source areas, especially when the Strait of Gibraltar acted as a barrier. I consider much of North America to be, in human terms, an extension of the Eurasian Plain (Chapter 2). Once humans reached eastern Siberia, only the Bering Strait would have prevented access to this sector at certain times.

The Middle East has been the predominant terrestrial access channel from Africa into Eurasia but the alternative route via the Horn of Africa may have been a significant alternative at times (Lahr & Foley, 1994; Quintana-Murci *et al.*, 1999). The early colonisations of eastern Asia by populations ancestral to those defined as *H. erectus* around 1.9–1.8 Myr (Klein, 1999; Aguirre & Carbonell, 2001) and western Asia by *H. ergaster* with Mode 1 technology around 1.7 Myr (Gabunia *et al.*, 2000, 2001; Bar-Yosef & Belfer-Cohen, 2001) are in keeping with this view. A further expansion (or expansions) of hominds with Mode 2 technology, appears to have reached the Middle East around 1.5–1.4 Myr (Aguirre & Carbonell, 2001; Belmaker *et al.*, 2002) and north-west Africa by 1 Myr (Raynal *et al.*, 2001). A subsequent colonisation around 800–500 kyr (probably 600 kyr) by hominids with Mode 2 technology, via the Middle East (Bar-Yosef & Belfer-Cohen, 2001), appears to have reached China (Hou *et al.*, 2000) as well as Europe (Aguirre & Carbonell, 2001) and may have included passage across the Strait of Gibraltar (Alimen 1975; Giles Pacheco & Santiago Pérez, 1987). Another possible expansion around 250–200 kyr, by a population claimed to belong to *H. helmei* with Levallois technology, could have reached Europe via the Middle East (Foley & Lahr, 1997; Porat *et al.*, 2002) and may have originated the Neanderthal line. A further expansion after 100 kyr, this time by Modern Humans, first spread eastwards across the tropical Asian belt and led to the colonisation of Australia before 50 kyr (Thorne *et al.*, 1999; Bowler *et al.*, 2003), the colonisation of Europe by at least 45 kyr (Klein, 1999) and the subsequent expansion across the Eurasian Plain, including for the first time North America. These are just some expansion markers in what would have been a more fluid and continuous system of range expansion and contraction over the last 2 Myr (Finlayson *et al.*, 2000a).

Populations that reached geographical areas away from tropical Africa would have differentiated and adapted to local conditions. The success of adaptation would have depended on the climatic and environmental stability of each area. More stable areas would have permitted populations to persist and reach densities close to carrying capacity. Populations with more restricted geographic ranges would have been most prone to extinction whereas those with greater mobility and behavioural diversity persisted (Potts, 1996a, b, 1998). *Homo erectus* may have survived until 25 kyr in parts of tropical South-east Asia (Swisher *et al.*, 1996). These would be the areas least affected by cold and aridity, probably more so than even tropical Africa (Pope, 1983). It is possible that the *H. helmei* expansion (Foley & Lahr, 1997) did not reach South-east Asia because of the vicissitudes of climate. If they did reach they may have been unable to establish themselves because of competition from established local populations, a theme that was probably recurrent in human dispersal throughout the Pleistocene.

Populations in mid-latitudes and on the Eurasian Plain would have survived by expanding the range northwards during warm events and contracting south during cold arid ones (Finlayson *et al.*, 2000a; Hewitt, 2000). Some of these populations may have been present when the next populations of colonisers arrived. It is possible that populations attributed to *H. heidelbergensis* and *H. helmei* would have been present in Europe at the same time (Foley & Lahr, 1997) and this was also the case between Modern Humans and Neanderthals. Increasingly, frequent cold and arid periods removed these populations from Europe and western Asia. The western European populations were constrained by the Strait of Gibraltar and repeatedly became extinct, a situation experienced by many other animals (Busack, 1986). The south-eastern European, or south-west Asian, populations of Neanderthals were apparently able to expand the range south into the Middle East (Tchernov, 1998) but the Sahara would have prevented further expansion to the south. It was only in South-east Asia that tropical refugia allowed persistence outside Africa. The populations that settled in Europe and north-western and central Asia could only maintain core geographical areas in the south (Gamble, 1999), across the topographically heterogeneous mid-latitude belt from Iberia to the Caucasus and beyond to the Altai (Figure 2.1). As such they evolved adaptations to a different way of life from that of the plains. These populations lived in areas where several ecological zones were in close proximity (Soffer, 1994) and where warm temperate climates permitted a varied dietary subsistence that resembled, though less varied, that in the tropics. Such conditions spread north during warm events and the humans tracked these. The extinction of the Neanderthals towards the end of Oxygen Isotope Stage (OIS) 3, when conditions were not as severe as earlier in the late Pleistocene, was the product of a series of cold and arid events which had the combined effect of suppressing and fragmenting Neanderthal populations (Chapter 7). Such periods of population contraction (bottlenecks) and expansions might also have had an effect on the genetic divergence of isolated lineages (Lahr & Foley, 1998, Hewitt, 1996, 1999, 2000). Open plains hunting strategies were, as a rule, evolved in the African plains. Such strategies pre-adapted African populations to exploit the structurally similar conditions of the Eurasian Plain, provided they could get there across the mountainous mid-latitude belt. Once established there, they could persist until extreme cold events forced them south into refugia.

The history of human demography and geographical dispersion can therefore ultimately be explained in terms of climate change causing vegetation change in turn causing change in the distribution and abundance of animals, particularly mammalian herbivores (Chapter 2). Very rarely is a direct climate–human population effect expected. Rather, it is the effect of climate on available habitats and resources that is the mechanism producing change. The question of human history in the late Pleistocene has a strong spatio-temporal, multi-scale, component in which events are nested within others at different scales that may even appear to be operating in reverse directions. Such situations would have created spatio-temporal mosaics in the distribution and abundance of human populations which are closer to expectations from ecological and evolutionary theory than the currently available views of a single or small number of discrete events (dispersals/migrations) that are proposed for the origins of 'modern' humans and their dispersal across the globe. To such mosaics must be added the strong cumulative temporal components that are often disregarded in issues of human origins.

Populations of humans therefore colonised areas of the Old World from Africa as from 1.9 Myr BP. A major contributing factor to this expansion was the habitual exploitation of an increasingly carnivorous diet (Gamble, 1995; Lahr & Foley, 1998, Stanford, 1999; Stanford & Bunn, 2001). On reaching mid-temperate latitudes of the Northern Hemisphere these humans were in areas that had the highest net primary productivity after regions near the Equator, although today, for example, the productivity peak is confined to the period from April to September (Field *et al.*, 1998). Other populations (e.g. *heidelbergensis*) would have followed (Lahr & Foley, 1998). Moderns and Neanderthals may have been separated by 500 kyr (Stringer, 1989) or less, if there was a subsequent expansion from Africa of humans with Mode 3 technology around 250 kyr (Foley & Lahr, 1997). Successive isolations, caused by the fluctuating climates of the Pleistocene, differentiated these populations genetically and phenotypically although the timescales involved were in all probability insufficient for speciation to have occurred. This means that there would have been gene flow between populations at certain times (Lahr & Foley, 1998) and these populations may have held fairly stable hybrid zones (Cain, 1971; Hewitt, 1996, 1999, 2000). The cumulative effects of successive isolations built on the genetic differences. A similar, but more limited because of the timescales involved, genetic differentiation after the Last Glacial Maximum (LGM) between Moderns in different parts of the Earth followed the earlier pattern.

Earlier in the late Pleistocene one of these forms increasingly became a large mammal hunter of the East African plains. Its geographical range expanded as savannahs and grasslands gained over tropical rainforests during cold and arid phases (Foley, 1987; Foley & Lee, 1989; de Menocal, 1995). During one or more milder and wetter phases the Saharan barrier broke down and their range extended into the Middle East. At such times the pressure would have been for northward range expansion into these new areas of grassland and savannah as existing areas to the south returned to forest. Similar expansions may have occurred during the cool conditions around 500–450 kyr, when *H. heidelbergensis* with Mode 2 (Acheulian) technology reached Europe (Klein, 1995; Lahr & Foley, 1998) and at the end of another cold period around 250 kyr when humans (attributed to *H. helmei*) with Mode 3 (Levallois) technology spread (Foley & Lahr, 1997; Lahr & Foley, 1998).

The mountains of Turkey and the Caucasus acted, as we have seen, as formidable barriers to range expansion but eastward dispersal was not impeded by barriers. Once across the western Asian mountains, African plains people found habitats in the Eurasian Plain structurally analogous to their East African original habitats and they were able to rapidly colonise. As these environments expanded with increasing cold and aridity, plains-adapted humans followed (Woillard, 1978; Suc & Zagwijn, 1983; Gamble, 1986, 1999; Roebroeks *et al.*, 1992; Zagwijn, 1992). This explains, in my view, why Moderns expanded. It is not expansion that 'occurs against the grain of climatic change' (Lahr & Foley, 1998). In Europe the local form, the Neanderthal, exploited highly heterogeneous landscapes principally in the south. These were the only areas that could sustain populations throughout a glacial–interglacial cycle (Gamble, 1999). During warm interglacials, the Neanderthals' range expanded northwards reaching its highest during, or just prior to and just after (when dense forests had not established themselves, Roebroeks *et al.*, 1992; Gamble, 1999), the last interglacial and coinciding with the establishment of the Mousterian technology (Foley & Lahr, 1997). The brevity of such events and gene flow with contiguous populations to the south precluded adaptation to the North Eurasian Plains environments, especially as much of the landscape would have been wooded (Roebroeks *et al.*, 1992; Gamble, 1999). The progressive cooling after the last interglacial gradually reduced the range of the Neanderthals.

The European case

The number of hominid dispersals from Africa into Europe that led to successful colonisations into Europe varied, depending on the author, between a single ancient event (Wolpoff, 1989), through two events involving *H. ergaster* and

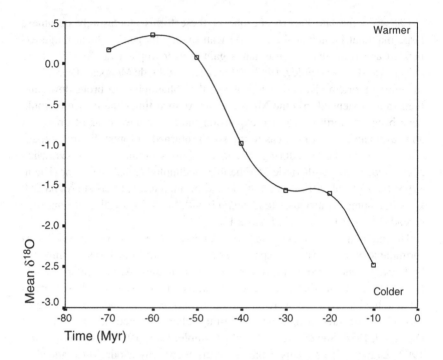

Figure 3.2. Decrease in temperature during the last 70 Myr. Regression model $y = -2.3029 + 0.0284x + 0.0028x^2 + (2.6 \times 10 - 5x)^3$. The relationship is statistically significant ($R^2 = 0.955$, $P = 0.016$). After Finlayson (2003).

H. sapiens (Klein, 1999) to up to four involving *H. ergaster*, *H. heidelbergensis*, *H. helmei* and *H. sapiens* (Foley & Lahr, 1997). In the latter case, Foley & Lahr (1997) have linked the four events to the introduction of technological modes 1 to 4 respectively. The fate of these populations on reaching Europe is also unclear. Although there are claims for the presence of *Homo* in Europe before 1 Myr (Martinez Navarro, 1997; Oms *et al.*, 2000), the earliest well-documented fossils are those of the TD6 level of Gran Dolina in Atapuerca (Spain) and the Ceprano (Italy) specimen which are older than 780 kyr (Parés & Pérez-Gonzales, 1995, 1999; Falgueres *et al.*, 1999; Manzi *et al.*, 2001). The Gran Dolina specimens have been given specific status – *H. antecessor* (Bermudez de Castro *et al.*, 1997) – and are considered to be ancestral to the Neanderthals and *H. sapiens*. Other authors consider that *H. antecessor* was a dead end and that the middle Pleistocene *H. heidelbergensis* was the ancestor instead (Hublin, 1998). I have discussed the difficulty of establishing the real situation earlier in this chapter. The fate of the Neanderthals is the most controversial of these events and one in which I will focus in Chapter 7.

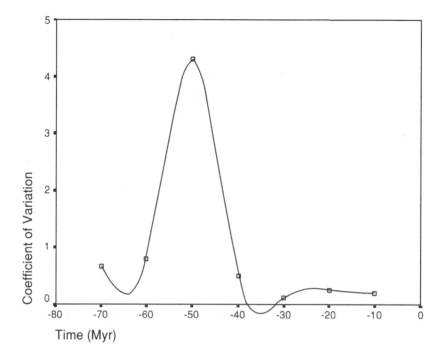

Figure 3.3. Climate variability pattern during the last 70 Myr. The relationship is statistically insignificant. After Finlayson (2003).

In a paper published in 2003 (Finlayson, 2003), I attempted to correlate the global climatic record during the past 800 kyr to the presence of hominids in Europe. The Mediterranean Basin stood out as a buffer region between the tropical African geographical core areas of *Homo* and the marginal regions of northern and central Eurasia. The following time periods were studied at different scales.

The period 70–0 Myr

The period 70–0 Myr (Miller *et al.*, 1987; Berggren *et al.*, 1995; Denton, 1999) was analysed in 10 Myr periods at a resolution of 1-myr intervals. The period was one of climatic cooling (Figure 3.2) and the last interval (10–0 Myr) was the coolest of the entire sequence. Climate variability (Figure 3.3) reached a maximum between 60 and 50 Myr and stabilised after 40 Myr at this scale. Although variable, the climate in the 60–50 Myr interval was significantly warmer than at any later stage.

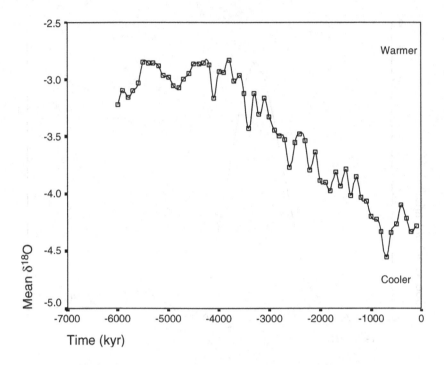

Figure 3.4. Decrease in temperature during the last 6 Myr. Regression model y =
$-4.3118 + 7.3 \times 10 - 7x + 1.9 \times 10 - 7 \times 2 \,(2.6 \times 10 - 11x)^3$. The relationship is
statistically significant ($R^2 = 0.955$, $P = 0.016$). After Finlayson (2003).

The period 6–0 Myr

The period 6–0 Myr (Denton, 1999) was analysed in 1-Myr periods at a resolution of 100 kyr intervals. There was a statistically significant trend towards
climatic cooling (Figure 3.4) and the last interval (1–0 Myr) was the coolest
of the entire sequence. Climate variability (Figure 3.5) also increased significantly throughout the period. The interval 1–0 Myr was therefore the coolest
and most unstable of the sequence, followed by the 2–1 Myr and 3–2 Myr
intervals respectively.

The period 850–0 kyr

The period 850–0 kyr (Ruddiman *et al.*,1986) was analysed in 50-kyr periods at
a resolution of 5 kyr intervals. There was a weak but significant trend towards

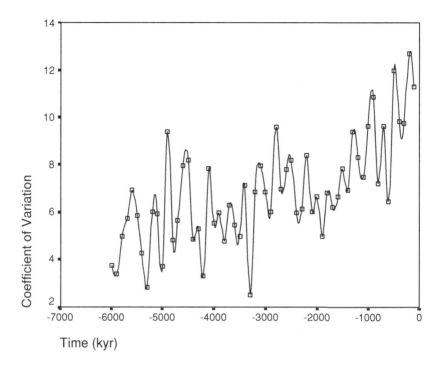

Figure 3.5. Climate variability pattern during the last 6 Myr. Regression model y = 11.862 + 0.0044x − 1.2 × 10 − 6 × 2 + (1.1 × 10 − 10)³. The relationship is statistically significant ($R^2 = 0.538$, $P < 0.0001$). After Finlayson (2003).

climate cooling (Figure 3.6). Climate variability appeared erratic at this scale and there was no significant trend (Figure 3.7) although the most unstable periods were in the latter half of the sequence. When mean temperature and variability were combined (Table 3.1) it became clear that the last 200 kyr were the coolest and most unstable of the sequence and that there was a trend from warm/stable to cool/unstable from 850–0 kyr.

The period 90–0 kyr

The period 90–0 kyr (GRIP, 1993) was analysed in 5 kyr periods at a resolution of 0.5 kyr intervals. There was no significant trend during this period (Figure 3.8) but there were two clear intervals of cooling, around 75–55 kyr, that is related to OIS 4, and from 45–15 kyr that is related to OIS 3 and 2. There was no trend in

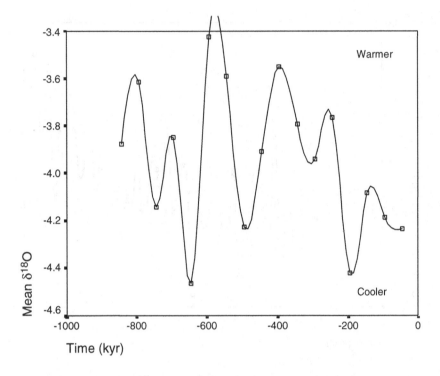

Figure 3.6. Relationship of temperature through time during the last 850 kyr. The relationship is statistically insignificant. After Finlayson (2003).

climate variability either (Figure 3.9), but there were unstable intervals clearly related to the cooling events, at 80–65 kyr, 50–25 kyr and 20–10 kyr.

Table 3.1 summarises the presence in Europe of members of the genus *Homo* between 850 and 0 kyr which may be related to the data in Figures 3.2–3.9. The following periods were identified by Finlayson (2003).

The period 1.7 Myr–850 kyr

There were insufficient data for analysis. The presence of *H. ergaster* in Dmanisi (Georgia) (Gabunia *et al.*, 2000), however, indicates that hominids had entered Eurasia by then. A number of European sites claim human arrival on the basis of presence of Mode 1 technology (Carbonell *et al.*, 1999a; Oms *et al.*, 2000) and it seems very likely that hominids were present in southern Europe around 1 Myr.

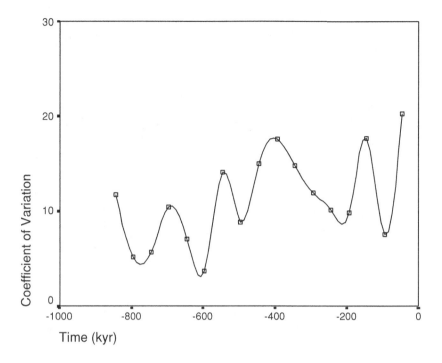

Figure 3.7. Climate variability pattern during the last 850 kyr. The relationship is statistically insignificant. After Finlayson (2003).

The period 850–600 kyr

This period is marked by the presence in southern Europe of fossils that are older than 780 kyr and may be ascribed to the form *antecessor* (Bermudez de Castro *et al.*, 1997; Manzi *et al.*, 2001). These hominids appear to use Mode 1 technology (Carbonell *et al.*, 1999a).

The period 600–250 kyr

This is marked by the presence in southern, western and north-western Europe of fossils that are ascribed to *H. heidelbergensis* (Klein, 1999). These hominids use Mode 2 technology. The later ones may be considered pre-Neanderthal (Arsuaga *et al.*, 1993).

Table 3.1. *Summary of main climatic episodes described in the text and the major human events during the past 850 thousand years (kyr)*

Time period (kyr)	Climate	OIS	Duration (kyr) of favourable (+)/ unfavourable (−) OIS	% of period	Potential Homo pattern	Homo type	Mode
850–800	wu	22–21	20 (+)	40	(Persistence)	?	1
800–750	**WS**	**21**	**50 (+)**	**100**	**Colonisation**	***antecessor***	**1**
750–700	**CS**	**20–18**	**37 (−)**	**74**	**Extinction [52%]**	?	
700–650	ws	18–16	30 (+)	60	(Colonisation/ Persistence)		
650–600	**CS**	**16–15**	**30 (−)**	**60**	**Extinction [100%]**		
600–550	**WS**	**15–14**	**35 (+)**	**70**	**Colonisation**	***heidelbergensis***	**2**
550–500	WU	14–13	46 (+)	92	Persistence	*heidelbergensis*	2
500–450	**CS**	**13–12**	**28 (−)**	**56**	**Extinction [35%]**	**? *heidelbergensis***	**2**
450–400	wU	12–11	23 (+)	46	Persistence	*heidelbergensis*	2
400–350	WU	11–10	38 (+)	76	Persistence	*heidelbergensis*	2
350–300	Wu	10–8	36 (+)	72	Persistence	*heidelbergensis*	2
300–250	wu	8	0 (+)	0	Persistence	*heidelbergensis*	2
250–200	**Ws**	**8–7**	**45 (+)**	**90**	**Colonisation/ Persistence**	***heidelbergensis/ helmei***	**3**
200–150	Cs	7–6	36 (−)	72	(Extinction) [22%]	? *neanderthalensis*	3
150–100	CU	6–5	28 (+)	56	(Persistence/ Extinction)	*neanderthalensis*	3
100–50	**CS**	**5–3**	**12 (−)**	**24**	**Extinction [43%]**	***neanderthalensis***	**3**
50–0	CU	3–1	11 (−)	22	(Persistence/ Extinction) [100%]	*neanderthalensis/ sapiens*	4

'Climate' column: w, warm; c, cold; s, stable; u, unstable; and capitals refer to major episodes. OIS, oxygen isotope stage. 'Potential Homo pattern' column: percentages behind extinction events are the duration of the extinction event divided by the duration of the preceding favourable event × 100. Data in bold: these are considered by author to be most significant and are highlighted in the text.

Source: After Finlayson (2003)

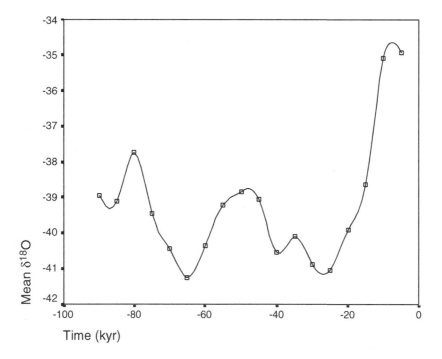

Figure 3.8. Relationship of temperature through time during the last 90 kyr. Regression model $y = -32.754 + 0.4679x + 0.0087x^2 + (4.7 \times 10 - 5x)^3$. The relationship is statistically significant ($R^2 = 0.678$, $P = 0.001$). After Finlayson (2003).

The period 250–150 kyr

This is marked by the presence of pre-Neanderthal fossils and the use of Mode 3 technology. Foley & Lahr (1997) have proposed that Mode 3 arrives from Africa with a new species – *H. helmei*.

The period 150–50 kyr

This is marked by the presence of the 'classic' Neanderthals (Klein, 1999) with Mode 3 technology.

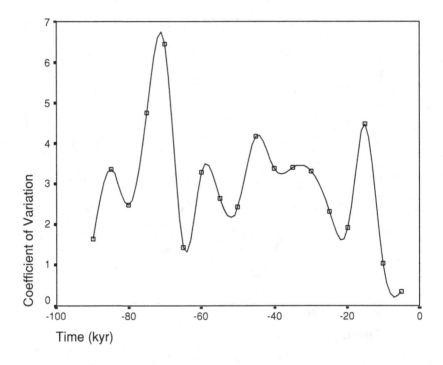

Figure 3.9. Climate variability pattern during the last 90 kyr. Regression model ln(y) = 1.3493 + (11.1463/x). The relationship is statistically significant (R^2 = 0.599, $P < 0.0001$). After Finlayson (2003).

The period 50–0 kyr

This is marked by the presence of the last Neanderthals and their extinction between 40 and 30 kyr (Klein, 1999). The Neanderthals continue with Mode 3 technology but, towards the end, appear to adopt (Mellars, 1999) or invent (d'Errico *et al.*, 1998) Mode 4-like technology (Chapter 5). The period is also marked by the arrival of Moderns and Mode 4 technology.

There has therefore been a consistent global cooling from 70–0 Myr. The last 1 Myr, the period of the colonisation of Europe by hominids, has been the coolest and most variable of all. The temperature and variability pattern within the last 850 kyr appears closely related to the pattern of hominid distribution (Table 3.1). Finlayson (2003) divided the last 800-kyr period into 50-kyr intervals and allocated to each the probability of colonisation of Europe from Africa or extinction of *Homo* populations in Europe.

Colonisation events

Mithen & Reed (2002), using computer simulations, also found that the highest probabilities of colonisations from Africa were before 1.7 Myr. Three post-1 Myr colonisation events were predicted by Finlayson (2003), all during the only warm and stable periods in the sequence.

800–750 kyr

This interval reflects the warm and stable conditions of OIS 21 that lasted for the entire 50 kyr period and coincided with the presence in Europe of the first undisputed hominids (*H. antecessor*) at Atapuerca (Pares & Pérez Gonzales, 1995, 1999; Falgueres *et al.*, 1999; Carbonell *et al.*, 1999b) and Ceprano (Manzi *et al.*, 2001) with Mode 1 technology. The earlier presence of hominids over 1 Myr in Europe was not discarded and would be related to warm and stable conditions preceding the climatic oscillations that commenced around 850 kyr with OIS 22.

600–550 kyr

This interval reflects the warm and stable conditions of OIS 15 that lasted for 35 kyr. The predicted colonisation events coincide with the presence in Europe of hominids attributed to *H. heidelbergensis* and the arrival of Mode 2 technology.

250–200 kyr

This interval reflects the warm and stable conditions of OIS 7 that lasted for 45 kyr. This predicted colonisation event coincided with the arrival of Mode 3 technology and, according to Foley & Lahr (1997), *H. helmei*. The last colonisation, by Moderns with Mode 4 technology, is not predicted and is discussed below.

Extinction events

Four extinction events were predicted. These were the four cold and stable events in the sequence.

750–700 kyr

This interval reflects the cold and stable conditions of OIS 20–18 that lasted for 37 kyr.

650–600 kyr

This interval reflects the cold and stable conditions of OIS 16 that lasted for 30 kyr. It seems highly unlikely that the *H. antecessor* populations that occupied the southern fringe of Europe would have survived two such major events so that my view is that these populations were probably not ancestral to later European populations of *Homo*.

500–450 kyr

This interval reflects the cold and stable conditions of OIS 12 that lasted 28 kyr. The fossil evidence is of insufficient resolution to support or reject this prediction. Given, however, that this event was of relative short duration and that it followed a lengthy period of colonisation and persistence, it is possible that the *H. heidelbergensis* populations with Mode 2 that had colonised north-western Europe were large enough to survive the crisis in southern European refugia. The subsequent presence of hominids attributed to *H. heidelbergensis* in Europe (Klein, 1999; Manzi *et al.*, 2001) would seem to bear this out. The fact that there is no predicted colonisation event after this, and therefore no new population that would account for the European fossils, would seem to lend further support to the persistence of *H. heidelbergensis* past the OIS 12 crisis.

100–50 kyr

This interval reflects the cold and stable conditions of OIS 4 that lasted for 12 kyr. This extinction event is slightly earlier than the extinction of the Neanderthals (40–30 kyr). The short duration of this event was clearly insufficient to cause the Neanderthal extinction but may have depressed the populations to such a level that they became extinct shortly after (Finlayson *et al.* 2000a).

Intervening periods

There were five intervening periods between predicted colonisation and extinction events (Table 3.1):

700–650 kyr

This interval coincides with OIS 17 that lasted 30 kyr, was weakly warm and stable and was sandwiched between two major cold events. Assuming the *H. antecessor* populations became extinct in the preceding stage then there is a slight possibility of a small colonisation during this period. Alternatively, this stage could have supported surviving *H. antecessor* populations that then went extinct in OIS 16. Thirdly, it is possible that there were no hominids in Europe at this time. These options must remain open on present evidence.

550–500 kyr

This interval coincides with OIS 13 that lasted 46 kyr and was a warm but unstable period. It was therefore a long period, following the colonisation by *H. heidelbergensis* with Mode 2 technology, that would have allowed the persistence of these populations.

450–250 kyr

This interval coincides with OIS 11–9 that lasted for 97 kyr and was also a warm but unstable period. Its length would have permitted the persistence of *H. heidelbergensis* populations that survived OIS 12.

200–100 kyr

This interval coincides with OIS 7–5 that lasted for 64 kyr. This was a cold period, the first part weakly stable and the latter unstable allowing for brief milder intervening periods. The preceding long interval of warmth of 250 kyr that may have included a new African colonisation and the arrival of Mode 3 technology seems to have permitted the persistence through the short 36-kyr interval that included the first part of OIS 6. The warm but short OIS 5 may have allowed a brief recovery of a population that is then identifiable as Neanderthal.

50–0 kyr

This interval coincides with OIS 2 that lasted 11 kyr. This was a cold and unstable period that saw the extinction of the Neanderthals, whose population had already been critically stressed in the preceding periods, and the severe population fragmentation and subsequent recovery in OIS 1 of Modern Humans (Demars, 1996; Torroni, *et al.*, 1998; Bocquet-Appel & Demars, 2000a).

These results indicate that there have been five major episodes that affected Europe:

(1)   The colonisation by *H. antecessor* or its ancestor with Mode 1 before 780 kyr which was restricted to the south. This population probably went extinct in either OIS 20, 18 or 16.

(2)   The colonisation by *H. heidelbergensis* with Mode 2 during OIS 15 between 600 and 550 kyr. This population survived for up to 400 kyr and the Neanderthals probably evolved directly from them.

(3)   A colonisation during OIS 7 between 250 and 200 kyr. This may have been in the form of hominids (*H. helmei*) carrying Mode 3, in which case their relationship with European *H. heidelbergensis* would be of great interest, or it may have been a technological diffusion from Africa at a time when the European and African populations were not isolated.

(4)   The Neanderthal extinction during OIS 3 between 40 and 30 kyr.

(5)   The survival and expansion of *H. sapiens* after 40 kyr.

Contact between African and European populations

There is therefore the distinct possibility that African and European populations met, at least, on two or three occasions with its clear implications for gene flow. The first would have been when *H. antecessor* or its ancestor colonised southern Europe and met earlier populations that may have arrived before 1 Myr. Alternatively, *H. antecessor* may have evolved from these populations in Europe. Further evidence of the pre-800 kyr colonisation of Europe is required. The second was the interchange between 250 and 200 kyr when Mode 3 was introduced into Europe. The question of whether the Neanderthals evolved from local *H. heidelbergensis*, from an African newcomer (*H. helmei*) or derived genes from the two populations is of great interest. Thirdly, there is the best-documented case of the arrival of Modern Humans around 40 kyr (Hammer, *et al.*, 1997; Malaspina *et al.*, 1998; Semino *et al.*, 2000; Shen *et al.*, 2000; Barbujani & Bertorelle, 2001; Reich *et al.*, 2001; Marth *et al.*, 2003) and their relationship to the Neanderthals that will occupy most of the discussion of this book.

Persistence of hominid populations

Foley (1994) showed that hominid extinction in the last 5 Myr was correlated with low temperature. The results presented here indicate an increasing capacity for persistence in Europe within the last 800 kyr. *Homo antecessor* was restricted to the south and probably went extinct. *Homo heidelbergensis* reached mild oceanic north-west Europe and persisted significantly longer. The Neanderthals colonised areas further east than *H.heidelbergensis* and persisted through conditions that were colder and more unstable than those experienced by *H. heidelbergensis*. Finally, Modern Humans arrived in conditions that would not have previously led to European colonisation. Their survival of OIS 2 was, however, not exceptional.

Geographical origin of early European Modern Humans

The entry of Moderns into Europe during the Upper Palaeolithic may have been in the form of a series of waves (Richards *et al.*, 2000; Semino *et al.*, 2000). The evidence that Moderns with Aurignacian technology (the earliest Mode 4; Klein, 1999) came from Africa is unconvincing (Sokal *et al.*, 1997; Richards *et al.*, 1998; Bar-Yosef, 2000). Otte (1994) has proposed that the Modern Aurignacian colonisation of the European Plains came from the eastern plains of Eurasia and recent genetic evidence supports this view (Semino *et al.*, 2000; Bosch *et al.*, 2001; Wells *et al.*, 2001). The early Palaeolithic expansion into northwest Siberia, prior to the split into western and eastern groups (Derbeneva *et al.*, 2002), would be consistent with a west-central Asian origin as would the very early appearance of humans in the European Russian Arctic at 40 kyr (Pavlov *et al.*, 2001) and in north-east Asia at 42 kyr (Brantingham *et al.*, 2001). In this scenario, the Aurignacians would have been people that colonized Eurasia during an earlier colonization event and developed the Aurignacian outside Africa. These populations may have been related to those that appeared in the Middle East around 100 kyr (Chapter 4) or they may have been part of the Mode 3 colonisation at 250–200 kyr or of a more diffuse dispersal from the Middle East to proximal regions between 450 and 200 kyr. Those populations colonising Europe would have interacted with *H. heidelbergensis* leading to the Neanderthals and those reaching *H. heidelbergensis*-free central Asian areas would have continued a separate evolution with close affinities with the Zagros Mountains to the south (Davis & Ranov, 1999) and colonising ecotonal zones in southern Siberia, to the north-east, by 130 kyr (Goebel, 1999). Thus the Neanderthals and the Moderns would share a recent common ancestor, as proposed by Foley & Lahr (1997), but the Neanderthal difference would have come from adaptation to local conditions (Finlayson *et al.*, 2000a) and/or hybridisation between *H. heidelbergensis* and the form named *H. helmei* by Foley & Lahr (1997). These populations would have only been isolated from each other for a maximum of 350 kyr.

The Mediterranean and hominids in the Pleistocene

The Mediterranean lies in an intermediate latitude between the African tropics and Europe. In the west, the Strait of Gibraltar is generally assumed to have been a barrier to human dispersal (Straus, 2001). In the east, the Levant is generally regarded a corridor and a point of contact (Tchernov, 1992, 1998).

The Middle East

The proximity of the Levant to East Africa and its relatively low latitude means that a greater number of human dispersals from Africa would be expected to reach that zone than Europe itself. If we work on the premise that unstable intervals (that would have included warm periods within), in addition to the warm stable ones that allowed colonisation of Europe, could have opened a window for humans to reach the Middle East then we predict that humans could have reached the region on at least six occasions. These would be: (a) 850–800 kyr; (b) 700–650 kyr; (c) 600–500 kyr; (d) 450–200 kyr; (e) 150–100 kyr; and (f) 50–0 kyr. Of particular interest is the extended period of 250 kyr between 450 and 200 kyr that could have permitted a number of dispersals of which only that with Mode 3 technology between 250 and 200 kyr reached Europe. The Middle East could have acted as a secondary source for populations to temporarily colonise adjacent geographical areas without penetrating into Europe. From here populations could have dispersed eastwards keeping to similar latitudes. Central Asian areas would also have been close and the presence of *H. ergaster* in Dmanisi (Gabunia *et al.*, 2000) indicates that penetration through the valleys of the Caucasus was possible from an early date. The other interesting period is 150–100 kyr that coincides with the presence of Moderns in the Levant that does not appear to have materialised into a European colonisation (Bar-Yosef, 1998). The subsequent presence of Neanderthals in these sites would coincide with the cold and stable conditions between 100 and 50 kyr that would have, according to the data presented here, caused a contraction of the Modern range. Finally, the 50–0 kyr period would have been favourable for a Middle Eastern colonisation.

The Maghreb

The Maghreb is in a similar situation to the Levant and similar periods of colonisation could have brought humans to this region. The barrier would have been the Sahara during cold and arid events (deMenocal, 1995). It follows that gene flow between Maghreb and tropical African populations would have been more frequent than between tropical African and European populations and, as with the Levant, there would have been a relatively continuous trickle of populations except in the coldest and most arid moments, i.e. 750–700 kyr, 650–600 kyr, 500–450 kyr, 200–150 kyr and 100–50 kyr coinciding with OIS 20, 18, 16, 12, 6 and 4.

The Iberian Peninsula

If the Strait of Gibraltar was a barrier for humans from 850–0 kyr then the Iberian Peninsula would always have been reached by colonisers that first reached central Europe. In that scenario, Iberia would be a part of the process of European colonisation and a refugium for stressed populations during cold events. The Strait of Gibraltar existed throughout this period although there were times when its width was considerably narrower than the present-day 14 km (Alimen 1975; Giles Pacheco & Santiago Pérez, 1987). Conditions of almost neutral current that flow between the Atlantic and the Mediterranean would have existed between glacials and interglacials (Fa *et al.*, 2001). The possibility that humans got across the Strait during this long period is likely. Populations able to cross would meet similar bioclimates and could colonise much of Iberia without first having to adapt to central European conditions. If we assume that the most favourable moments would have been between warm and cold events, as populations became compacted through habitat shrinkage and current flows were favourable, then the following periods would be predicted as likely: around 750 kyr, 650 kyr, 500 kyr and 200 kyr. To this we may add another moment prior to the time scale under investigation that would have permitted an entry of humans that were the precursors of *H. antecessor* (? *H. ergaster*). The third event would account for the similarity in Acheulian artefacts on both shores of the Strait of Gibraltar (Alimen, 1975; Giles & Santiago Pérez, 1987) and the fourth for the presence of Mousterian in Iberia and the Maghreb (Allsworth-Jones, 1993). The absence of potential crossings after this would explain why the Aterian never reached Iberia (Debénath, 2000; Straus, 2001, but see Bouzouggar *et al.*, 2002).

Italy and the Balkans

Crossings from Africa into Italy would have been less likely (Alimen, 1975). The Italian (Mussi, 1999) and Balkan (Bar-Yosef, 2000) peninsulas may have acted as short-term refugia during glaciations but their smaller area relative to Iberia and their more continental disposition would have rendered them less effective.

The role of the mid-latitude belt

The mid-latitude belt that, as we saw in Chapter 2, stretches from Portugal and Morocco in the west to the Himalayas in the east (Figure 2.1), has not previously been considered a unit in biogeographical terms. Yet, in terms of faunal composition, topography and vegetation structure (Chapter 2) it stands as a single unit, especially in the centre and west. In addition, the northern areas (not dissected from the north by the Mediterranean) were the most important glacial refugia (Chapter 7). In my view understanding this mid-latitude belt as a unit is crucial in understanding human evolution in Eurasia and North Africa. The colonisation of this belt by an early population that diverged from the African hominids around 500 kyr (Chapter 5) is central to our understanding of the Neanderthals and other archaic hominids. These robust archaic hominids, often referred to as *heidelbergensis*, would have been well-adapted to the topography and fauna of this belt. Westwards and eastwards dispersal, probably from an origin close to the entry point from Africa that is most likely to have been in the Middle East, and further adaptation to the conditions of this wide area led to the evolution of the Neanderthals. Isolation in glacial refugia may have differentiated Iberian, Balkan and Caucasus populations but gene flow during milder intervals would have maintained cohesion which allows us to lump them all as Neanderthals. I predict that, as more data become available, differences between Iberian, Balkan, Caucasus and other Neanderthal populations will become apparent. There would have been a parallel spread along the southern shores of the Mediterranean to north-west Africa but these populations would have remained isolated from the Neanderthals by the Mediterranean and also by the periodic expansions of the Sahara. This would explain the similarities and differences between the Jebel Irhoud humans and the Neanderthals (Hublin, 1992; Allsworth-Jones, 1993), but see Simmons & Smith (1991) and Smith *et al.* (1995) for an alternative perspective that requires contact across the Strait of Gibraltar. A similar separation of eastward-expanding populations along the southern mid-latitude belt would not have produced similar differentiation as there would have been no physical barrier comparable to the Mediterranean or the Sahara. As we have seen, the North African and Middle Eastern populations would have been subsequently more influenced than the Neanderthals by African gene flow. The situation may have been similar in the Upper Palaeolithic even though the expanding populations in Europe were arriving across the plains and not the mid-latitude belt itself. The Modern North African Berber mtDNA lineage appears to have diverged from the most ancient European and Middle Eastern lineage around 50 kyr, one-third of their genes having a Middle Eastern ancestry and one-eighth being sub-Saharan (Macaulay *et al.*, 1999). North-west African populations had a 75% Y chromosome contribution to the gene pool from East Africa contrasting with a 78% contribution to the Iberian population from western Asia along the northern rim of the Mediterranean with the Strait of Gibraltar acting as a strong, but incomplete, barrier (Bosch *et al.*, 2001). These results further emphasise the western/central Asian genetic input to western populations of Moderns, being more pronounced in Europeans than in North Africans that continued to receive a proportional contribution from Africa.

Synthesis

In Chapter 2, I identified four main elements that contributed to human evolution and expansion: (1) medium-sized mammalian herbivore availability; (2) terrain topography; (3) habitat and landscape structure and distribution; and (4) physical and ecological barriers. From this chapter we can add the following refinements:

(1) As increasing climate variability created increasing spatial heterogeneity of habitats and landscapes in Africa, hominids became increasingly dependent on intermediate and open habitats and landscapes, evolving adaptations that permitted them to deal with the spatial uncertainties characteristic of such environments. Such adaptations, which may be summarised as adaptations that increased the scale of activity and improved environmental resistance, incidentally enabled hominids to expand their

geographical range away from the tropics, surmounting previous ecological barriers, with increasing chances of success.

(2) The climatic variability of the Pleistocene produced repeating scenarios of changing habitats, landscapes and barriers that enabled African hominids to expand the geographical range, or shift it northwards, and colonise Eurasia on multiple occasions as they tracked these changing environments. This climatic variability also created situations, especially marked in northern areas such as Europe, of high extinction probability of regional hominid populations. As geographical range expanded, the probability of populations becoming isolated when climate introduced barriers, would have increased. In the case of African and Eurasian populations there would have been instances of genetic isolation of populations but the predominant theme would have been one of gene flow particularly among geographically proximate populations.

(3) In this chapter I have established four potential time periods of colonisation from Africa: pre-1.7 Myr, that may have involved a number of colonisation events – 800–750 kyr (OIS 21), 600–550 kyr (OIS 15) and 250–200 kyr (OIS 7). These need not be strict colonisations and may instead involve the spread of cultural innovations. We have also established five further periods during which gene flow would have existed between Africa and Eurasia. These periods could also have received African hominids or cultural innovations but largely to south-west Asia (from where they could have subsequently dispersed). These periods were: 700–650 kyr (OIS 17); 550–500 kyr (OIS 13); 450–250 kyr (OIS 11–9); 200–100 (OIS 7–5); and 50–0 kyr (OIS 2–1). In the latter case I predict such events only during the warmer OIS 1 (10–0 kyr). Finally, four periods when non-tropical Eurasian extinctions were most likely were identified: 750–700 kyr (OIS 20–18); 650–600 kyr (OIS 16); 500–450 kyr (OIS 12); and 100–50 kyr (OIS 4).

4 *The Modern Human–Neanderthal problem*

Current theories of Modern Human origins are divisible into two groups. There are those that promote regional continuity and hybridisation and those that advocate a recent African origin to all Moderns (Klein, 1999). In the first category is the strict Regional Continuity Model (Wolpoff, 1989) which proposes that ancestral populations of an archaic hominid dispersed from Africa across the Old World around 1.9 Myr and that the populations that settled in different parts of the world independently evolved into Moderns. For this to have happened, without the different populations becoming distinct species, the model predicts that there was regular gene flow between populations. Subsidiaries of the Regional Continuity Model have been advanced. Brauer (1992) proposed that there was a degree of regional continuity between populations but that there was a significant African genetic contribution to European and western Asian populations through hybridisation and assimilation. Smith (1992) proposed a similar model but reduced the importance of the African contribution with a smaller number of genes being assimilated by European and western Asian populations. The 'intermediate' models would seem to have some support from the genetic evidence (Templeton, 2002).

The late Pleistocene Out-of-Africa Model (Cann *et al.*, 1987; Stringer & Andrews, 1988) is the parent of the rival group. It proposes that Moderns evolved in Africa between 130 and 200 kyr ago, spread out of Africa and replaced all other archaic human populations after 100 kyr ago. There would therefore be no genetic contribution from any archaic group (e.g. the Neanderthals) to the Modern Human gene pool. A variant is the Weak Garden of Eden Model (Harpending *et al.*, 1993; Sherry *et al.*, 1994; Ambrose, 1998). It differs from the 'classic' Out-of-Africa in that it proposes that there is no population increase after an initial expansion from Africa around 100 kyr and a major demographic expansion between 70 and 40 kyr. The populations that established themselves in different regions of the Old World were small and widely dispersed and suffered genetic bottlenecks (Haigh & Maynard Smith, 1972; Jones & Rouhani, 1986; Harpending *et al.*, 1993; Sherry *et al.*, 1994; Rogers & Jorde, 1995). There was a subsequent population expansion of these genetically isolated populations between 70 and 50 kyr, which was related to new technologies (Upper Palaeolithic/Late Stone Age) that increased the

71

environmental carrying capacity for human populations. The Multiple Dispersals, Bottlenecks and Replacement Model attempts to provide a mechanism for ex-African expansions. This model sees the environment as the driving force (Lahr & Foley, 1994, 1998; Foley & Lahr, 1997) and the Middle to Upper Palaeolithic technological transition as a key factor. According to this model the ancestral African population was reduced in size, experiencing a genetic bottleneck, due to climate-driven habitat fragmentation. A series of population increases with dispersal followed by further bottlenecks characterised human expansion. According to Lahr & Foley (1998) the ancestor of the Neanderthals, that they name *Homo helmei*, would have dispersed within and out of Africa during Oxygen Isotope Stage (OIS) 7 or 8. The first dispersal, into the Middle East, occurred during the mild OIS 5. Subsequent cooling caused a population retreat, these Modern Humans being presumed not to be behaviourally or physiologically adapted (or at least not as well as the contemporary Neanderthals) to the cold of Eurasia. A second dispersal, in OIS 4 or early OIS 3, enabled the dispersal of a population into Asia and a final one, around 45 kyr, coinciding with the Middle–Upper Palaeolithic transition, into the Middle East was rapidly followed by the colonisation of Europe. This model recognises that, as Moderns spread they replaced archaic populations including the Neanderthals. Lahr & Foley (1998) attempt to provide a mechanism that is based on existing theoretical frameworks of evolutionary ecology and biogeography. They also recognise that evolutionary events, such as Modern Human origins, have a strong geographical component and highlight the vital link between demography and spatial distribution.

The alternatives available to us until now have therefore required that the fate of archaic groups is determined either by 'eviction' or 'continuity/replacement' (Lahr & Foley, 1998; Tattersall & Schwartz, 2000). An alternative hypothesis has recently been proposed that does not require the intervention of Moderns in the extinction of the archaics which is seen as part of a natural and recurring process of habitat fragmentation during glacial cycles that severely affected non-tropical hominid populations (Finlayson, 1999; Finlayson *et al.*, 2000a). This model, which is developed in this book, differs from the traditional and, until now apparently mutually exclusive, alternatives of replacement (usually by competition) or continuity that do not consider non-human related extinction of archaic populations, including the Neanderthals, to be important. Patterns of hominid evolution and the key elements of Modern Human behaviour can be explained within the framework of the general principles of evolutionary ecology (Foley, 1992). This alternative model is precisely based on theoretical evolutionary ecology and geography (Hutchinson, 1959; MacArthur & Wilson, 1967; MacArthur, 1984; Brown, 1995).

The use of culture as an all embracing and all pervading explanation to the evolution of Moderns has obscured the processes by which Moderns evolved (Foley, 1989). Throughout this book I view humans as components of ecological communities with the driving force behind change being natural selection acting to 'keep up' with the spatio-temporal heterogeneities of Pleistocene Earth. As these heterogeneities became more marked so populations that were adapted to cope with change fared best. Behavioural attributes that permitted rapid adjustments to change were selected. Humans increasingly became refined risk managers.

The species problem

Before discussing the biology of Neanderthals and Modern Humans we should establish who they were and what their relationship to each other was. The debate concerning Modern Human origins often seems to revolve around whether or not followers of a particular camp regard the two to be distinct species or not. The point about the definition of Neanderthals and Moderns, or indeed any other human, is that it is a taxonomic concept. The discussion about human origins must be an evolutionary one and, whether or not we are advocates of cladistics, taxonomy should only be seen as a convenient tool in packaging and not as a proxy for evolutionary thinking. In evolutionary terms it does not matter what we call Neanderthals or Moderns. The point is that the genetic evidence, which is the only reliable tool that we have today, indicates that Neanderthals and Moderns had a common ancestry that can be approximately dated at around 500–400 kyr and that the two lineages apparently went along separate paths, one in Eurasia and the other in Africa. Physical barriers, aided by climate, apparently kept the two lineages apart until they re-met in Eurasia some time after 100 kyr ago (depending on location). We presume they did not have genetic contact in the interim but it is only a presumption. As we saw in the previous chapter it is a presumption that is unlikely to have held across the entire geographical range throughout the period 500–40 kyr. What happened when the two lineages met? We cannot be certain because the evidence is so meagre. One thing seems clear from the genetic evidence – no Neanderthal genes survived (Krings *et al.*, 1997, 1999, 2000; Ovchinnikov *et al.*, 2000). We should not be surprised if at some point in the future contrasting evidence is found. Why could Neanderthals and Moderns not interbreed and leave mixed traits? There is no biological reason whatsoever but clearly, on present evidence, the Neanderthal genetic contribution is nil and may have at best been very small. There is no reason either to suspect a uniform pattern across space. Put in simple terms – what happened in France need not have been the same as what happened in Java.

The origins of humans (members of the genus *Homo*) date approximately to the Plio-Pleistocene boundary around 2 Myr (Wood & Collard, 1999; Hawks *et al.*, 2000). Whether we choose to consider *H. erectus* and *H. sapiens* to be a single, evolving, species (Hawks *et al.*, 2000; Wolpoff & Caspari, 2000) or separate species (Stringer, 2002a; Tattersall, 2002) does not alter the nature of the discussion of this book that is concerned with the evolutionary ecology of populations and not with taxonomic definitions. For the purpose of this book it is enough to recognise a speciation event, probably subsequently unparalleled, somewhere near the Plio-Pleistocene boundary that led to the first member of the genus *Homo* (Mayr, 1950; Wolpoff & Caspari, 2000). Thereafter, we lack the resolution to allow precision in the identification of lineages as the Pleistocene picture is likely to have been so complicated spatially and temporally. In this context we should note that, in North American songbirds at least, the paradigm that many species originated as a consequence of the late Pleistocene glaciations has been shown to be flawed. Instead the glaciations were an ecological obstacle through which only some species were able to persist (Klicka & Zink, 1997). There is no doubt that, among humans, there would have been many cases of geographical separation leading to the emergence of distinct populations. Where isolation was sufficiently long the trajectories, as in the case of the Neanderthals, would have led to distinct morphological and related features. This, on its own, does not make the Neanderthals a distinct species as some authors seem to suggest (Lieberman *et al.*, 2002). Whether such differences were of a kind that precluded interbreeding when populations met once more, thus confirming the presence of distinct biological species (Cain, 1971), is something that we cannot answer today. In any case we have to be aware that reproductive isolation even between good species may, in some cases, be imperfect (Schluter & Nagel, 1995). Morphological distinctness, the basis for allocating fossils into species, is only a general, and not infallible, guide in the delimitation of species (Simpson, 1951; Cain, 1971). The weakness of relying on morphology is especially evident if we consider the phenotypic plasticity of most organisms (Geist, 1998). Genetic differences are, equally, subject to our own protocols and definitions. The splitting of the *Homo* phylogeny is therefore subjective and not directly relevant to the question of Modern Human origins and the extinction of archaic populations.

So there would have been multiple branches in the evolution of *Homo* in the Pleistocene, especially as the geographical range expanded and the chances of isolation became greater. There must also have been continuity in at least one population, that which led to the Moderns. We may therefore best regard modern *H. sapiens* to be the terminator, for now, of an ancestral-descendant sequence of interbreeding populations that evolved independently of others – the gens described by Simpson (1951). A number of authors have attempted to link the emergence of Moderns to a speciation event (Crow, 2002). The evidence in favour is inconclusive. A study of a highly variable sub-terminal non-coding region from human chromosome 16p13.3 did not reveal a signal for population growth in Africa that would be expected if such a speciation event had taken place (Alonso & Armour, 2001).

Evidence of widespread hybridisation between Neanderthals and Moderns would certainly be suggestive but so far we only have the claim from Lagar Velho in Portugal (Duarte *et al.*, 1999; Zilhao & Trinkaus, 2002), based on morphology, and that is it. This recent discovery of a skeleton in Portugal, claimed to be a Modern–Neanderthal hybrid and dated at 25 kyr (Duarte *et al.*, 1999), is in apparent conflict with the genetic evidence. The skeleton was found buried in a distinctively Upper Palaeolithic pattern, implying behavioural modernity, but its anatomy was claimed to be a mosaic of Neanderthal and early Modern Human features. The claim has been vigorously contested by some who feel that the skeleton lacks distinctive Neanderthal features (Tattersall & Schwartz, 1999). Duarte *et al.* (1999) and Zilhao & Trinkaus (2002) claim the skeleton as evidence in support of interbreeding between early Moderns and Neanderthals. The authors recognised the inappropriateness of applying a strict biological species distinction, based on reproductive isolation, to Neanderthals and early Moderns. They also rejected hypotheses of full replacement of late archaic humans by early Moderns everywhere outside Africa and instead saw the need for an approach that brought together regional complexities, temporal, human biological and cultural processes as well as the historical trajectories that took place. If this child was a hybrid, then the claim for widespread hybridisation between Neanderthals and Moderns rests on the dating evidence that suggests that the hybrid was in existence up to 5 kyr after the extinction of the last Neanderthal in Portugal. We have to accept that transposing what happened in a single valley to the whole world is risky. In the Middle East, Neanderthals and Moderns supposedly occupied the same geographical area for longer than anywhere else (Arensburg & Belfer-Cohen, 1998). It is an area that has produced fossils of Neanderthals and Moderns but so far no hybrids. So we cannot, presently, use the biological species concept to determine whether we are dealing with one or two species. There is something that is even more worrying and for that I must now turn to the question of convergent and parallel evolution.

I have already said that the only available solid evidence that we can draw upon is the genetic evidence. The reason is that I seriously question the validity of arguments based solely on morphological comparisons. The problem is exacerbated by the small sample sizes available, which oblige researchers to combine specimens from distant parts of the geographical range and from different time periods, often making statistically unsatisfactory inferences. Throughout the animal kingdom we find numerous cases of unrelated species converging biologically in response to similar ecological problems (Cody, 1974, 1975). The point is that the probability of convergence in response to similar pressures has to, logically, be even greater among closely related forms because they are starting off from templates that are quite similar to each other. This means that we need genetic evidence to support any evolutionary conclusions that we draw from morphology because genes tell us a history that is independent. So can we differentiate, especially when we only have single or even small groups of specimens, between lineages and convergence on morphology alone? The answer is that we cannot.

At any point during the late Pleistocene, Neanderthals, Moderns and other contemporary human populations are best regarded as a *sapiens* polytypic species (Cain, 1971; Aguirre, 1994; Smith *et al.*, 1995). A time slice at a point in the late Pleistocene would reveal a range of human populations spread across parts of Africa, Eurasia and Oceania. Some would have been genetically linked to each other, behaving as sub-species, while the more extreme populations may well have behaved as good species with minimal or no inter-breeding. The two extremes were probably in operation at different times and in different parts of the world. The human array at any point should best be regarded as a polytypic species of common descent and varying degrees of subsequent isolation. This view is in keeping with the increasing evidence that demonstrates that species across their range are often divided into patchworks of parapatric sub-species and races with intervening hybrid zones (Hewitt, 1989).

When a species is separated by a geographic barrier and the terminal forms gradually diverge and eventually behave as two distinct species when they meet on the other side we have an example of a polytypic species that is known as a 'ring species' (Cain, 1971). For Mayr (1942) such 'circular overlaps' perfectly demonstrated the process of speciation. It is likely that the varying levels of isolation, gene flow and distance among human populations in the Pleistocene generated geographical distribution patterns at particular times that were akin to the ring species concept. For this reason it will be useful to explore this concept, and its most recent developments in particular – sympatric and parapatric speciation – a little further. In particular, I focus on the effects of gene flow in preventing speciation. The Out-of-Africa vs Multiregional debate focuses on whether there was isolation or gene flow between Pleistocene human populations (Hublin, 1998; Hawks & Wolpoff, 2001). Genetic exchange undoubtedly slows down the rate of divergence of two populations (Irwin *et al.*, 2001; Porter & Johnson, 2002) but a more pertinent question is by how much? Recent speciation models have highlighted the importance of local adaptation as a process that can oppose gene flow leading to rapid population divergence to the level of full species (Rice & Hostert, 1993; Johannesson, 2001). Even in cases of complete sympatry, strong selection can eliminate gene flow between populations leading to very rapid speciation (Gavrilets *et al.*, 1998; Kondrashov & Kondrashov, 1999). Even though sympatric speciation is likely to be rare it appears a distinct possibility in competitor-free, resource-diverse, environments (Dieckmann & Doebeli, 1999; Filchak *et al.*, 2000; Wilson *et al.*, 2000) and minor changes in the selective environment can cause population divergence (Danley *et al.*, 2000). According to Gavrilets *et al.* (1998), rapid speciation is also possible without the need for extreme founder effects, complete geographical isolation or selection for local adaptation. Short-term reductions in migration rate were sufficient to produce significant and irreversible divergence and reproductive isolation in just several hundred generations. Divergent selection pressures between populations can also lead to divergent sexually selected traits, if these are favoured in different environments (Endler, 1992; Schluter & Price, 1993; Schluter & Nagel, 1995; Irwin, 2000; Irwin *et al.*, 2001; Johannesson, 2001). Development, by providing a context for cryptic divergence in the allelic basis of regulatory interactions and creating interspecific incompatibilities, also increases the probability of speciation even in cases of strong gene flow (Porter & Johnson, 2002). At the other end of the scale we have the classic allopatric speciation models in which geographically isolated populations can diverge due to genetic drift even in the absence of strong divergent selective pressures but this process will be severely curtailed in the presence of migration.

Irwin *et al.* (2001), in their review, concluded that the role of gene flow in preventing differentiation of the terminal forms of a ring species should be highly dependent on whether initial substitutions were favoured everywhere or only in parts of the species range. We can at least conclude that demonstration of gene flow in the case of sympatric or parapatric Pleistocene human populations does not automatically preclude lineage divergence, or indeed even speciation. Given the differences in spatial ecology between Neanderthals and Moderns, that will become apparent in this book, we should not be surprised to observe lineage separation in the presence of varying degrees of gene flow as detected by Templeton (2002).

Sympatry or allopatry?

The situation that arose in Europe and western Asia that concluded with the extinction of the Neanderthals and the colonisation of the Moderns was not exceptional, as we saw in the previous chapter. The pattern of extinction of Neanderthals does not follow an east to west gradient as would be expected if the Moderns arriving from the Middle East had replaced them. Instead, Neanderthals became extinct across the mid-latitude belt from Portugal to the Caucasus at about the same time (31–29 kyr) (Finlayson, 1999; Finlayson *et al.*, 2000a; Ovchinnikov *et al.*, 2000; Zilhao, 1996; Smith *et al.*, 1999; Chapter 7). Populations that had occupied areas to the north, around the North Eurasian Plain, became extinct earlier (by 40 kyr). This, together with the long-established contemporaneity of Neanderthals and Modern Humans in the Middle East for thousands of years (Bar-Yosef, 1998) questions the long-held view that Moderns caused the Neanderthal extinction. The dating of a Javan specimen, attributed to *H. erectus*, at 25 kyr (Swisher *et al.*, 1996) indicates a late persistence of archaic humans also in tropical South-east Asia. Since we now know that Moderns had reached well into Australia by 50 kyr (Thorne *et al.*, 1999; Bowler *et al.*, 2003), protracted geographical overlap between Moderns and archaics must have been widespread. Questions that relate to reproductive, ecological and behavioural interactions in areas of geographic overlap (sympatry) therefore assume a greater relevance. Because the European–Middle Eastern region is the best documented, it is issues of Modern Human–Neanderthal interactions that are receiving prominence. Sympatry would have been possible if Neanderthals and Moderns had sufficiently different niches to permit ecological isolation (Lack, 1971; Cody, 1974) or if numbers were such that populations were below carrying capacity. Competition would only occur in situations in which the populations were at carrying capacity and resources became limiting. Differences in ecology may explain the long periods of sympatry (Mellars, 1996). Recent work suggests that Moderns and Neanderthals were ecologically separated and had distinct habitat preferences (Finlayson, 1999; Finlayson & Giles Pacheco, 2000). Improved resolution of climatic data is allowing greater precision in linking ecological change with human behaviour (van Andel & Tzedakis, 1998). The rapid changes during the late Pleistocene (Allen *et al.*, 1999) especially in zones of sharp ecological transition (Peteet, 2000) have clear implications for the survival of populations, including hominids. The importance of temperate and tropical refugia is also being re-assessed and isolation in cold-stage refugia (e.g. Iberia, southern Italy, Balkans) is reflected in distinctive present-day patterns of genetic variation and subdivision among widely different animals (Willis & Whittaker, 2000). The evidence increasingly points to the Modern expansion and the Neanderthal extinction being the products of habitat and resource change during the late Pleistocene, with southern refugia playing a critical role in the outcome (Finlayson, 1999; Finlayson & Giles Pacheco, 2000). The degree of interaction between Moderns and Neanderthals would have been minimised by ecological separation. Contact would be predicted to be greatest where heterogeneous landscapes were close to the plains and would therefore have been localised. So far the only case of apparent hybridisation, as we have seen, is the Lagar Velho child (Duarte *et al.*, 1999; Zilhao & Trinkaus, 2002). The key is not whether hybridisation occurred but its effect on the human gene pool. Given the available genetic evidence (Krings *et al.*, 1997, 1999, 2000; Ovchinnikov *et al.*, 2000; Caramelli *et al.*, 2003) it would seem that such hybridisation must, at best, have been restricted to localised hybrid zones (Hewitt, 1989). In the same way, the conditions required for competition (Finlayson *et al.*, 2000b) would not appear to hold given the low population densities (Mussi & Roebroeks, 1996; Harpending *et al.*, 1993) resulting from the constantly and rapidly changing climate (GRIP, 1993; Allen *et al.*, 1999). Competition, like hybridisation, may have been a very local phenomenon with no consequence to the Neanderthal extinction. It would be very informative to have ecological data from South-east Asia where late *H. erectus* and *H. sapiens* must have been sympatric for at least 25 kyr.

Genes

Studies of mitochondrial (mtDNA) and fossil (fDNA) Neanderthal DNA (Krings *et al.*, 1997, 1999, 2000; Ovchinnikov *et al.*, 2000; Scholz *et al.*, 2000) indicate their genetic distinctness when compared to present-day humans. We lack, however, a comparison with Modern Humans that were contemporary with the Neanderthals (Wolpoff, 1998) although a recent comparison with 24 kyr-old Modern Humans indicates a genetic discontinuity (Caramelli *et al.*, 2003). In any case these observations do not exclude the Multiregional model (Nordborg, 1998; Reletheford, 1999). The time of the last common ancestor of Modern Humans and Neanderthals is now put within the time frame of 317–741 kyr, possibly around 465 kyr (Krings *et al.*, 1997, 1999; Ovchinnikov *et al.*, 2000). From the limited data available the provisional conclusion that may be drawn about Neanderthal genetic diversity is that it was low, comparable to Moderns, and much lower than for the great apes. Since Neanderthals had a larger geographical range than the apes, it appears that the Neanderthals may have expanded from a small population (Krings *et al.*, 2000). If so, it would seem that Neanderthals were similar to Moderns in demographic expansion characteristics, low mtDNA and nuclear diversity in Moderns being equated to a rapid population expansion from a small population (Jorde *et al.*, 1998).

Many genetic studies in the 1980s and 1990s seemingly clarified the question of a single African origin (between 100 and 200 kyr) and the timing of genetic differentiation of human populations around 100 kyr (Cann *et al.*, 1987; Vigilant *et al.*, 1991). However, not all molecular clocks tick at the same rate (Strauss, 1999) and there may even be variations in rate through time within the same lineage. A number of studies now propose faster mutation rates than conventionally accepted (Siguroardottir *et al.*, 2000). Effects include a more recent placing of the time of mitochondrial 'Eve' and of major Pleistocene human population expansions (Excoffier & Schneider, 1999). A study of the haplotypes of the PDHA1 gene (that apparently has a steady mutation rate) on the X chromosome threw the dating of Modern Human origins and the issue of a single African origin wide open. Ingman *et al.* (2000), however, tested and confirmed that human mtDNA lineages evolved at constant rates. Only the D-loop did not evolve at a constant rate and was therefore unsuitable for dating evolutionary events.

Harris & Hey (1999) found a fixed DNA sequence difference between African and non-African samples and the age of onset of population subdivision was around 200 kyr. This evidence supported earlier studies (Harding *et al.*, 1997; Hammer *et al.*, 1998) that pointed to Asian ancestry older than 200 kyr that was hard to reconcile with a unidirectional Out-of-Africa migration 100 kyr and the total replacement of archaic populations in Asia. This message was reinforced in another recent study (Reletheford & Jorde, 1999) that, while supporting a major role for Africa in Modern Human origins, left the question of complete African replacement open. In other words, it was not clear whether the gene pool of Moderns was completely African or predominantly so (Jorde *et al.*, 2000). Recent high resolution studies using the Y-chromosome and of complete mtDNA sequences appear to have strengthened the Out-of-Africa perspective further (Ingman *et al.*, 2000; Underhill *et al.*, 2000; Richards & Macaulay, 2001) but the question of complete replacement of all archaic human populations by Moderns is still in doubt (Templeton, 2002).

The evidence is also pointing toward multiple dispersals from Africa. A study of a 565-bp chromosome 21 region near the MXI gene, which is unaffected by recombination and recurrent mutation, and confirmed by independent evidence from a Y-chromosome phylogeny, suggests a series of distinctive range expansions: a first one to Oceania via South Asia; a second one to east Asia and subsequently north-east Asia and America; and a third mainly to Europe via west and central Asia (Jin *et al.*, 1999). This observation is consistent with the view that aboriginal Australians and some Asians, in addition to Africans, carry ancient DNA sequences (Harding *et al.*, 1997; Stoneking *et al.*, 1997; Kaessmann *et al.*, 1999). A population bottleneck appears to coincide with a Eurasian colonisation from Africa, estimated to have occurred at 38.5 kyr and no earlier than 79.5 kyr (Ingman *et al.*, 2000). These observations point to an early dispersal of Moderns into Asia via the Horn of Africa (Lahr & Foley, 1994; Foley, 1998; Quintana-Murci *et al.*, 1999; Kaessmann *et al.*, 1999) around 120–100 kyr, and a subsequent dispersal that included Europe between 60 and 40 kyr (Lahr & Foley, 1994; Underhill *et al.*, 2000). Both dispersals originated in eastern Africa (Quintana-Murci *et al.*, 1999).

The greater genetic diversity of African populations (Kaessmann *et al.*, 1999) has also been used as evidence of its greater age and, therefore, its function as source (Tishkoff *et al.*, 1996; Jorde *et al.*, 1997; Harpending & Rogers, 2000). Genetic diversity is not just a function of time (and distance from source) but also of effective population size (Ingman *et al.*, 2000) and the African population size was indeed larger than in other parts of the world during recent human evolution (Reletheford & Jorde, 1999).

The assumption that mtDNA is inherited by the maternal line alone has been challenged as it appeared that mtDNA from the mother's egg could recombine with sperm-contributed DNA (Awadalla *et al.*, 1999; Eyre-Walker *et al.*, 1999; Hagelberg *et al.*, 1999). This potentially set the clock out and even questioned the very existence of a mitochondrial 'Eve'. Estimates of relatedness could be affected because recombination would create a more genetically homogeneous population through time than would otherwise be predicted so that differences between more diverse ancient sequences and more homogeneous recent ones would be exaggerated (Strauss, 1999). The mtDNA recombination idea has been strongly challenged by Ingman *et al.* (2000) and Elson *et al.* (2001).

A recent study of 62 human population samples confirmed that the demography of populations strongly affected genetic affinities, those not undergoing demographic expansion showing increased genetic distances from other populations. Otherwise, genetic affinities closely matched geography (Excoffier & Schneider, 1999). The genetic study of population expansions (Excoffier & Schneider, 1999) may go some way towards focusing genetic research on recent human evolution away from the 'Out-of-Africa/Multiregional' debate with questions that will require multidisciplinary collaboration.

In a revision of the genetic evidence, Harpending & Rogers (2000) concluded that the evidence in support of the Out-of-Africa model was far less clear than it had been five years earlier. The issue of absence of evidence of a population expansion in a number of gene loci was a particular problem and this issue has been used by Hawks *et al.* (2000) to refute the Out-of-Africa model. For now, balancing the available evidence, we can conclude that Africa, probably East Africa, was the source area of Moderns. There were probably three major Out-of-Africa expansions: one around 1.9 Myr; another one around 840–420 kyr, that coincides with predictions of a demographic explosion around 500 kyr (Aguirre, 2000) shortly after the emergence of the early archaic Modern form in Africa (Brauer *et al.*, 1997); and a third around 150–80 kyr (Templeton, 2002). This third expansion may have involved an early phase, around 120–100 kyr, to Oceania via the Horn of Africa and South Asia and a second phase, around 60–40 kyr, to East Asia (and eventually North America) and also into West and Central Asia and from there into Europe. It is also likely that intermediate periods saw varying degrees of isolation by distance. The degree to which these dispersals involved complete replacement or, instead, some degree of inter-breeding remains unresolved. The divergence of the Neanderthal and Modern Human lineages around 465 kyr would be consistent with the second expansion described above. We should not discard, either, the possibility that population expansions were not always in the same direction and that 'reversals' would have taken place probably in response to sudden, opposing, climatic trends that were typical of the Pleistocene (Chapter 6).

Ecomorphology

There are many excellent texts that describe the morphological characteristics of Neanderthals and Moderns (see, for example, Klein, 1999). For this reason I will limit myself to highlighting and contrasting the major features of the two forms and to discussing their functional significance. The particular features of the Neanderthal morphology have been attributed to the gradual accretion of characters during long periods of isolation in Europe and western Asia (Hublin, 1998). Some of the features that characterised the Neanderthals were present in pre-Neanderthals (Arsuaga *et al.*, 1997; Lebel *et al.*, 2001) indicating that they had been evolving independently in their geographical area for a long time. Genetic drift is often considered a major factor in the evolution of the Nean-derthals' particular morphology even though there is no specific evidence to corroborate this assertion. On the other hand, a number of features are consid-ered adaptive and are thought to reflect the particular environments exploited by the Neanderthals. I will now summarise these features.

The Neanderthals were very robust, barrel-chested and exhibited muscular hypertrophy. The hand's morphology permitted a very powerful grip. Together with strong and cortically thick leg bones these features suggest the ability for endurance in use. The shafts of the phalanges of the foot may have been adapted for prolonged movement over irregular terrain (Klein, 1999). The low angle between the neck of the femur and the shaft is a characteristic of highly active individuals (Trinkaus, 1993). A number of morphological features are difficult to interpret and this has led some authors to suggest that such features had no functional significance (Klein, 1999). Some of these features may reflect the highly mobile lifestyle of the Neanderthals (Chapter 5). These relationships are, however, tentative. The longer and thinner pubis of Neanderthals when compared to Moderns (Rosenberg, 1988; Dean *et al.*,1986) may reflect a longer gestation period (Trinkaus, 1984) but this is disputed (Anderson, 1989; Rak, 1990; Stringer & Gamble, 1993). The precocity of Neanderthal children (Dean *et al.*, 1986; Trinkaus, 1986), along with a long gestation period, could reduce the need for females to remain in fixed locations for extended periods. I interpret these features in the context of the behavioural ecology of the Neanderthals (see Chapter 5). They are indicative of a form occupying heterogeneous landscapes with a behaviour pattern that required long and sustained movement and a close-quarter hunting strategy.

Of the cranial features that typify Neanderthals, the large incisors may be the easiest to explain. The evidence of high wear in these teeth suggests that they had a non-masticatory, clamp-type, function (Brace *et al.*, 1981; Smith, 1983; Trinkaus, 1993). The size and shape of the skull and the brain, reaching higher volumes than in Moderns (Stringer & Gamble, 1993), has not been satisfactorily explained although it is now clear that differences between Modern Humans and Neanderthals arise early in ontogeny (Ponce de Leon & Zollikofer, 2001). Cold climate explanations, as we will see below, are inadequate. The significance of the supra-orbital torus (brow ridge), attributed by some authors as a skeletal adaptation to localised and intermittent biomechanical stresses (Russell, 1985), is similarly not satisfactorily understood (Stringer & Gamble, 1993).

The large body mass, breadth of trunk and shortness of the limbs have been interpreted as adaptations to the cold (Ruff, 1994; Holliday, 1997a, b). Attempts to contrast lower limb length with climatic and mobility variables among extant human populations have been based on poor data sets and have been inconclusive (Holliday & Falsetti, 1995). The large nose is also regarded as an adaptation to the cold (see below). Body mass is closely correlated with a number of biological parameters including basal metabolic rate, respiratory and circulatory physiology, locomotion, growth and reproduction, social dominance, mating success, and size and types of food consumed. In addition phylogenetic trends are known to correlate with body size at the species level in mammals (Alroy, 1998). Surface area and metabolic rate are closely linked to body size so that changes in body size for reasons other than climate can affect changes in surface:volume ratios (Brown & Gibson, 1983; Peters, 1983). Body size in mammals is additionally linked with primary productivity (Rosenzweig, 1968), available prey size (McNab, 1971) and fasting endurance in seasonal environments (Lindstedt & Boyce, 1985). The difficulty of establishing causality is exacerbated in species that are capable of modifying their environment culturally. For example, Moderns developing systems of storage would effectively be reducing seasonality, and the selective pressure for fasting endurance would be reduced.

The perception of the Neanderthals as cold adapted to the conditions of glacial Europe is prevalent (Brose & Wolpoff, 1971; Trinkaus, 1981; Gamble, 1993; Stringer & Gamble, 1993). The evidence in support of this notion is, however, inconclusive. If it is argued that Neanderthals and Moderns were different species then intra-specific ecogeographical rules (Bergmann's and Allen's in particular) should not be applied (Mayr, 1963; Cain, 1971). Even if the two forms are regarded as belonging to the same species the validity of applying such rules to spatially and temporally separated populations is suspect given that it would be unreasonable to assume that between different geographical areas and time periods the only independent and changing variable affecting morphology was climate. In addition conclusions based on strong assumptions of uniformitarianism, which rely on analogy with extant organisms, need to be treated with caution (Van Valkenburgh, 1994). Morphology can determine the ecological range of the phenotype and may limit ecological and geographical distribution (Ricklefs & Miles, 1994). The environment also affects morphology in a range of ways including physical factors (e.g. climate), habitat and landscape structure (e.g. via locomotor structures), diet breadth and resource partitioning (Ricklefs & Miles, 1994). It is therefore very difficult, and even unrealistic in a multivariate world, to attempt to correlate morphology to a single variable such as climate (e.g. Dayan *et al.*, 1991) unless all these other variables are regarded to be constant among the forms being considered and across the vast stretches of their geographical ranges, which clearly they were not. The expectation is particularly unrealistic for such highly adaptable forms as the Neanderthals and Moderns that would not have been restricted to very specific conditions. To argue for cold adaptation when Neanderthals and their predecessors survived for tens of thousands of years in the variable climates of Europe, in which full glacials only took up a small part of the time (Chapter 6), and predominantly in southerly latitudes (Chpater 7), is illogical.

It could be argued that while such criticism is valid for post-cranial morphological features such as limb proportions, whose characteristics would have a clear ecological dimension, it might not be for other features, such as nose shape and size. The large protruding nose of Neanderthals has been considered to have been an adaptation to extreme cold and arid conditions, its function being to warm the air entering the lungs (see review in Stringer & Gamble, 1993) although others have suggested that the Neanderthal nose may have been a means of losing heat generated by a very active life style (Franciscus & Trinkaus, 1988). Czarnetzki (1995), in studying the Neanderthal nose, on the other hand, concluded that it must have been adapted to the peripheries of hot, humid regions, perhaps even subtropical to moderate biotopes. Possible experimental support for a warm climate function comes from studies of selective brain cooling in mammals (du Boulay *et al.*, 1998). These workers found that the internal carotid artery blood of macaques was 1–2 °C below the core body temperature and that selective brain cooling was a major factor constantly at work in a normal primate. The cooling occurs after the blood leaves the aorta and is due to the close proximity of the carotid artery to the trachea, larynx and pharynx. The dominant influence on the lowered temperature within the internal carotid artery in monkeys was apparently the temperature of inspired air. The anatomy is very similar to that of humans and these workers concluded that a similar mechanism is at work in humans.

Part of the observed neurocranial variation in contemporary Moderns is related to climate, thermoregulation having a greater effect on the cranium than upon the body as a whole (Beals *et al.*, 1984). For Neanderthals living in the warm conditions of mid-latitude Europe, and with presumably an even greater brain-cooling problem than Moderns on account of their larger brain volume (e.g. Stringer & Gamble, 1993), a nose with a reduced surface area exposed to air passing through the nasal cavities (Czarnetzki, 1995) could have acted as a tool for reducing the warming of inhaled air. This could confer a significant advantage to individuals that might have their hours of activity curtailed by the risk of heat strain (e.g. Ulijaszek, 2001). An air-warming function would instead increase the brain-cooling problem. As du Boulay *et al.* (1998) point out, heavy exercise or a hot and humid atmosphere with the consequent intake of warm air could further increase brain metabolic rate with the production of still more heat. The functional morphology of the facial characteristics of Neanderthals, with their inflated cheeks (Stringer & Gamble, 1993), would also seem to be in need of reassessment since it is apparent that brain cooling in humans is aided by cooled blood flowing from the facial skin backwards through the orbital veins towards the cavernous sinus, which may act as a heat exchanger (Kratzing & Cross, 1984).

The evidence available therefore runs counter to the notion that Neanderthals were morphologically and physiologically cold-adapted. Instead, the supposedly warm-adapted Moderns were able to fare better in these cold conditions, undoubtedly assisted by cultural devices. Some authors have gone to the extreme of regarding the Neanderthal morphology to be 'hyperarctic', drawing analogy with modern Eskimo populations (Ruff, 1994; Holliday, 1997a). Certainly, the Iberian populations of Neanderthals would never have been exposed to the climatic extremes of regions presently occupied by these northern peoples. If we analyse the mean annual temperature (T) variation across stations in a quadrant bound in the north by $70°$ N, in the south by $50°$ N, in the west by $150°$ E and in the east by $130°$ W (covering Alaska, eastern Siberia and adjacent areas) for the period 1950–1990 (Climate Research Unit, University of East Anglia), we find that the range of T falls between $-0.28°$C and $-2.23°$C (mean $-1.33°$C). Comparable data for the Iberian Peninsula gives a range of $16.77°$C to $18.23°$C (mean $17.48°$C). In Iberia, the ecological conditions suitable for Neanderthal survival would have disappeared with a decline in T of between $8°$C and $10°$C (Chapter 7), long before such low Arctic temperatures would have been reached. These results also put into question the validity of predictions of the temperatures to which Neanderthals were exposed based on extrapolations from morphology (Holliday, 1997b). It is similarly unlikely that

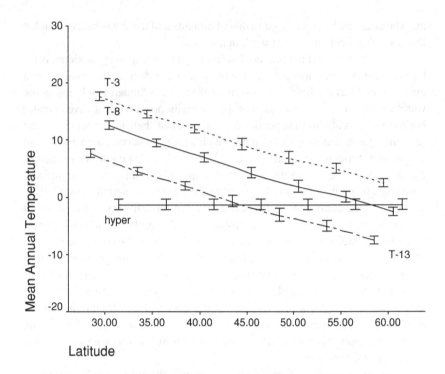

Figure 4.1. Estimated effect of drops in mean annual temperature (T) with latitude. The horizontal 'hyperarctic' line corresponds to a present-day T for Alaska and eastern Siberia. Since ecological conditions necessary for Neanderthal survival would have disappeared from, for example, Iberia with a drop in T of between 8 and 10 °C (Chapter 7), it is predicted that Neanderthals would not have experienced 'hyperarctic' conditions over the greater part of their geographical range.

Neanderthals further north in Europe would have ever been subjected to such low values of T, as their habitats would have similarly disappeared before such extreme temperatures were reached. In Figure 4.1 I show the results of extrapolating current T values, in latitude bands of 5° across a belt of Europe from 10° W to 10° E, by depressing T by 3 °C, 8 °C and 13 °C. The latter temperature would represent the extreme temperature drop at the height of the last glaciation (Dansgaard *et al.*, 1993). The present-day T range for the Alaskan–Siberian ('hyperarctic') quadrant is also included as a reference. It is clear from Figure 4.1 that the T drop at the height of the last glaciation would have produced values of T comparable to the hyperarctic sample in latitudes north of approximately 45°. Drops in T more in keeping with prevailing conditions in OIS 3, when the Neanderthals became extinct, would have produced hyperarctic conditions only further north, above the 55th parallel. It is highly unlikely that Neanderthals ever lived north of this parallel (approximately the latitude of Copenhagen, Denmark) and few would have lived north of the 45th parallel (approximately the latitude of Milan, Italy) in cold episodes. The conclusion is that Neanderthals would never have been exposed to T values remotely near those that could be considered hyperarctic. Whether or not Neanderthals had physiological and morphological characteristics that alleviated existence in cool and arid climates is not critical to questions of survival.

Cold-adaptation is therefore an inadequate explanation of Neanderthal morphology. Bone robusticity, shape and orientation of joints, and size, and orientation and mechanical leverage of their muscles are most likely to reflect an emphasis on strength and endurance required for high mobility, close-contact hunting with short-range weapons and production of high reaction forces in the upper limb (Churchill, 1998).

Moderns differed morphologically from Neanderthals in a number of ways. Obvious differences in cranial shape included a more flattened, retracted, face (Lieberman *et al.*, 2002) that may reflect decreased use of the anterior teeth (Klein, 1999). Skull shape may also reflect structural changes in the brain (Lieberman *et al.*, 2002). We cannot, however, distinguish at present between possible functional differences in cranial shape between Moderns and Neanderthals and differences due to genetic drift in small populations of these hominids. Cranial robusticity is linked to cranial size but not shape. Moderns from the late Pleistocene to the present have significant reductions in cranial dimensions, estimated at 10–30% between Upper Palaeolithic humans and recent Europeans. These changes have not been uniform and there are significant regional differences today. The Australian aboriginal pattern, however, bears no relationship to that of early Upper Palaeolithic hominids and is considered to reflect adaptation to Pleistocene Australia over a long period of isolation after an early entry (Lahr, 1994; Lahr & Wright, 1996). This entry may have occurred before the complete fixation of the mtDNA lineage now found in all living people (Adcock *et al.*, 2001).

The reduction in post-cranial robusticity and the relatively longer limb proportions in Moderns in all probability reflect adaptive responses. For reasons explained with reference to the Neanderthals, bioclimatic explanations are unsatisfactory. Reduction in robusticity, which continues in the Upper Palaeolithic and into the Holocene (Frayer, 1984), is probably related to a gradual reduction in habitual load levels (Trinkaus, 1997) related to the use of projectile technology and associated social correlates as well as to the nature of the landscape being exploited. Long distal limbs would have been particularly advantageous in the development of long-range mobility strategies over flat, open, terrain

(Chapter 5). Pelvic differences with Neanderthals may also reflect locomotory, rather than reproductive, differences (Klein, 1999).

Behaviour

Issues of behavioural complexity in middle Palaeolithic humans have come to the fore in recent years, taking over from the stagnation of fossil and archaeological discoveries (Stringer, 2002b). The discovery of 400 kyr-old wooden spears in Schoningen, Germany (Thieme, 1997), revealed the extent to which our perception of technological development, which has been based largely on stone tools, underestimated the capabilities of archaic humans. The possible existence of Lower and Middle Pleistocene huts and evidence of the early control of fire also point to the very early origin of technology that must have rapidly widened the niche of *Homo* (Gamble, 2001).

The links between technology and biology are increasingly diffuse (van Peer, 1998) as it appears that archaic *sapiens* had the capacity for 'modern' behaviour as reflected by the use of particular technologies in relation to changing ecology (Ronen, 1992; Foley & Lahr, 1997; McBrearty & Brooks, 2000). Technology also varied in response to environmental change among Moderns (Blades, 1999). Were changes in technology responsible for human range expansions or were they a response to new ecological settings? The recent confirmation of the antiquity (*c.* 50 kyr) of the Lake Mungo 3 (Australia) Modern Human skeleton (Thorne *et al.*, 1999; Bowler *et al.*, 2003) with evidence of the use of ochre (considered a modern trait) raises important questions concerning the geographical and temporal distribution of modern behaviour. These issues are addressed in detail in Chapter 5.

Patterns

The polarised human origins debate has, in seeking the ultimate solution, dampened the very nature of the mosaic that has been the spatio-temporal pattern of human evolution in the last two million years in response to the pressures imposed by climate change. From tropical African origins different forms of *Homo*, with different degrees of genetic distinctness, have populated the world with increasing success. Where populations have met, the outcome has not been dependent on the populations themselves but, invariably, on external factors. Whether there was genetic mixing or not would have depended on the degree of genetic distinctness and the probability of contact. Regional differentiation and continuity, replacement or colonisation of vacant areas are segments of a continuum. The use of culture as an all embracing and all pervading explanation to the evolution of Moderns has obscured the processes by which Moderns evolved (Foley, 1989). Patterns of hominid evolution and the key elements of Modern Human behaviour are explicable in terms of the general principles of evolutionary ecology (Foley, 1992).

The Out-of-Africa and multiregional perspectives are inadequate and have been cited as an example of a pre-paradigm scientific dispute (Strkalj, 2000). The Out-of-Africa postulate is not a formal theory but a bundle of inferences, supported by a range of evidence, while the multiregional perspective is a restatement of the single species model and is untestable (Foley, 1998). A central problem today is the absence of a coherent interdisciplinary approach and a unified theory of human evolution is urgently required. Such a theory needs to bring together the wealth of new information that is emerging from many different disciplines and make testable predictions that will give focus to future research. For it to be effective it must take the form of theory development, hypothesis testing and pattern seeking, and research must focus at multiple levels. I attempt to do this in this book.

Fossils cannot tell their own story and they need the comparative perspective of evolutionary biology. To progress towards a comprehensive and unified theory of human evolution we must have, as our base, a life sciences approach that links theory and mechanism (Foley, 2001). It must depart from the level at which selective processes operate – that of the individual. The confusion that is apparent in sectors of the palaeoanthropological community with regard to the roles of micro- and macro-evolutionary processes and the dismissal of the neo-Darwinian synthesis reveals a basic misunderstanding of the process of evolution (Tattersall, 2000; Foley, 2001). I am in general agreement with Foley's (2001) view. There is no need to invoke a special, non-Darwinian, mechanism to account for species diversity – in any case no satisfactory mechanism has been proposed. I also agree with Foley (2001) in his 'branching tree' view of human evolution. Where I differ is in my perception of the nature of the process. Branching and speciation are not inevitable consequences even if often they are the result. There are many examples of lineages that have not diverged, but have evolved or remained relatively unchanged over very long periods. I am not convinced either that all branchings automatically reflect speciation events. This is why I have, returning to Mayr's (1950) conclusion, stated in this chapter that there was a single, unparalleled speciation event, that led to the emergence of the *Homo* lineage. Between the view of a single, evolving, species and a tree full of species lies a view of diverging lineages, many of which do not make it to species status, extinctions and genetic intermixing. In my view we should be in agreement if we replace 'species diversity' by 'genetic diversity'. The evolution of *Homo* in the Pleistocene is marked by increases and declines in genetic diversity associated with demographic fluctuations, colonisation events and extinctions. If we view the world as one in which lineages continuously increase in diversity, we run the risk of seeing it as a world in which the trend towards diversity replaces other, antiquated, perspectives of directional evolution.

I agree with Foley's (2001) view of the role of demography and geography in providing the link between local and large-scale processes. To understand human evolution we must first understand the small-scale ecological processes in operation, next the macro-ecological ones, and finally the biogeographic patterns that reflect the outcomes of the smaller-scale processes. The problem lies in that we do not have the resolution to detect the small-scale patterns. We can partly resolve this by establishing larger-scale patterns and interpreting them within a theoretical framework of known multi-scale patterns and processes (Finlayson, 1999). The growth and shrinkage of populations are the result of the sum of the successes and failures of the individuals that form these populations. Where many populations are successful we have a successful species. The opposite is species extinction. The process of natural selection acting on individuals thus leads to multi-scale outcomes, from the success or demise of individuals, to that of local populations, regional populations and, ultimately, the permanence or extinction of species. In this book I am concerned, at one level, with the factors that led to the success and permanence (until now) of Moderns and the extinction of the Neanderthals. We may be looking at species-level patterns, in which case the loss of the Neanderthals would represent the extinction of a species, or lower-level patterns such as the extinction of the European populations of a form of *Homo*. Either way, understanding the processes at work is my goal. At another level we have the causes of the extinctions of the separate sub-populations that made up the European Neanderthal population. There may be many, small-scale, causes and these will be more difficult to discern, primarily because we lack the necessary archaeological, palaeontological and dating resolution. Too often issues of Modern success and Neanderthal extinction are pitched at this lower, largely unresolvable, level which may explain the sterile debate that has dominated the subject.

Synthesis

In this chapter we have commenced our focus on the Eurasian–African system that generated the Neanderthals and the Moderns that will be the focus for the rest of the book. We have identified the following features:

(1)   Human evolution need not be seen in the polarised 'multiregional' vs 'replacement' perspectives. The conclusions reached in Chapter 3 regarding the dynamics of colonisations, range shifts and extinctions form the basis

for a model that predicts a range of possible outcomes. It puts particular emphasis on extinctions and subsequent re-colonisations, the frequency of extinctions being related to environmental fluctuations, the abilities of hominid populations to persist and contingent demographic and spatial circumstances.

(2) Genetic evidence confirms some of the colonisation events predicted in Chapter 3. Three major expansions from Africa are predicted: around 1.9 Myr; 840–420 kyr (probably 500 kyr); and 150–80 kyr. The latter may be subdivided into three expansions: 120–100 kyr to South-east Asia and then Oceania via the Horn of Africa; 60–40 kyr to East Asia and then north-east Asia and North America via the Middle East; and 60–40 kyr to western and central Asia and then to Europe.

(3) With this evidence I have produced a revised table of African–Eurasian events over the last 2 Myr (Table 4.1). The following significant periods are identified:

   a. (Events 1–3) 2 Myr-0.95 myr covering many warm periods from OIS 71 to 25. There is at least one major colonisation involving *ergaster/erectus* that reaches eastern Asia, the Caucasus and possibly Iberia. There may have been several such events or instead subsequent cultural diffusion (including the expansion of Mode 2) and gene flow. Further clarification is required;

   b. (Events 5, 7) two periods of possible gene flow coinciding with OIS 21 and 17;

   c. (Events 9–10) a major colonisation event around 500 kyr, coinciding with OIS 15, involving *heidelbergensis* and subsequent gene flow during OIS 13. The Neanderthals would have evolved from this population as predicted by Arsuaga *et al.* (1997);

   d. (Events 12–14) a period of possible gene flow coinciding with OIS 11–9 followed by the arrival of Mode 3 technology in OIS 7 that appears to occur through cultural diffusion and gene flow and not a colonisation event. This is followed by the colonisation of south-west Asia by Moderns in OIS 5 and the subsequent expansion to South-east Asia and Australia;

   e. (Event 16) colonisation and diffusion events after 10 kyr associated with Neolithic expansions. Once again, the Modern expansions and the spread of Mode 4 are unrelated to the type of environmental conditions that would predict African movements. It is therefore most likely that the 60–40 kyr expansions are of Modern populations that had arrived in south-west Asia in OIS 5 and subsequently expanded from there. It also indicates that the early Upper Palaeolithic Eurasian technologies were not of African origin but instead reflected adaptations to plains environments in Eurasia.

Table 4.1. Predicted major expansions from Africa into Eurasia (dark grey rows). Pale grey rows are additional periods of potential gene flow. Note the predicted potential in the period 2–1 Myr. The period 1–0 Myr includes periods of potential isolation, range contraction and extinction (white). Event 16 refers to the Neolithic expansions in OIS 1. Note there is no other suitable moment for African colonisation in Event 16 which reinforces the view that the colonisation of Eurasia by Moderns was an Asian expansion. (See also text)

Event	Time period (Myr)	OIS	Colonisation	Cultural diffusion	Gene flow	Hominid	Technology
1	2.0–1.5	71–49	+	+	+	*ergaster/erectus*	1
2	1.5–1.0	47–27	+	+	+	*ergaster/erectus*	2
3	1.0–0.95	25			+		
4	0.95–0.8						
5	0.8–0.75	21			+		
6	0.75–0.7						
7	0.7–0.65	17			+		
8	0.65–0.6						
9	0.6–0.55	15	+		+	*heidelbergensis*	
10	0.55–0.5	13					
11	0.5–0.45						
12	0.45–0.25	11–9			+		
13	0.25–0.2	7		+	+		3
14	0.2–0.1	7–5	+			*early sapiens*	
15	0.1–0.05						
16	0.05(0.01)–0	1	+	+	+	*sapiens*	4/Neolithic

(4) Gene flow between African and Eurasian populations is predicted to have been possible for approximately 87.5% of the time in the last 2 Myr (Table 4.1). Gene flow appears possible for the entire period from 2 Myr to 1 Myr and for 75% of the period 1 Myr to the present. Lineage differentiation can occur in situations of gene flow. Thus prolonged periods of gene flow between African and Eurasian populations are not incompatible with the observed differentiation between Modern Humans and Neanderthals. The Modern Human and Neanderthal lineages split around 500 kyr, probably coinciding with the arrival in Europe of the ancestors of the Neanderthals, but the evidence required to justify their separation as distinct species is unavailable. Human populations across the globe at any point in the Quaternary are best regarded as forming a polytypic species with varying levels of genetic distinctness between each population.

(5) Modern Humans and Neanderthals were sympatric for long periods in certain areas, most notably the Middle East. A similar situation, requiring study, may have occurred in South, east and South-east Asia between Moderns and archaic '*erectus*' humans. The various forms within a geographical area appear to have behaved as ecotypes. Hybridisation in areas of geographical overlap is likely but supporting evidence is required.

(6) Modern Humans and Neanderthals exhibit very low genetic diversity which suggests that these populations went through significant bottlenecks during their evolutionary history. Such bottlenecks would have occurred during founding events or during periods of range contraction in southern refugia. Southern refugia, situated within the mid-latitude belt, were vital to the persistence of temperate Eurasian hominds.

(7) The robust Neanderthal and early Modern Human morphology is typical of middle Pleistocene hominids (Trinkaus & Rhoads, 1999). It reflects the continuing adaptation to the exploitation of mammalian herbivores in intermediate habitats and landscapes over heterogeneous terrain. Any attributable climatic advantage of the Neanderthal robust morphology (Pearson, 2000) must, in my view, be secondary. The gracilisation of Moderns reflects adaptation to more open environments over more homogeneous terrain and a change in hunting strategies with increased use of long-range projectiles. Humans have become increasingly refined risk managers. With increasing environmental variability, genetically-determined behavioural plasticity has been selected. In the rapidly changing world of the late Pleistocene, behavioural flexibility has been the mechanism that has permitted immediate response to abrupt environmental changes. This is the subject of Chapter 5.

5 Comparative behaviour and ecology of Neanderthals and Modern Humans

An understanding of the ecology of any species must include a knowledge of what it eats, where it finds it (and also water) and how it catches and processes it, where, when and with whom it breeds, where it obtains shelter and how it avoids predation and competition. These are problems common to all animals and need to be examined at different scales in order to fully comprehend them: daily, seasonal and inter-annual cycles may all have a bearing on a population's survival. Similarly, the spatial scale of operation of individuals (territories/habitats), groups (home ranges/landscapes), metapopulations (regions) and the species as a whole (geographical range) are critical in understanding its ecology. It follows that the patterns we may observe may be heavily dependent on the scale at which we observe them. In the case of humans one thing that will emerge throughout is that there are problems associated with generalisation at small scales. The world of Pleistocene humans, especially Neanderthals, has to be seen as a spatio-temporal mosaic at the scale of human generations. This makes it very difficult, as we will see, to establish generalised hypotheses other than at the large-scale, ultimate, levels of causality. I will now examine aspects of Neanderthal and Modern Human ecology from the perspective of resource acquisition with the view of comparing and contrasting the two forms.

Food and feeding ecology

Any comprehensive theory of hominid evolution must rest heavily on a theory of resource acquisition (Kaplan & Hill, 1992). In the specific case of the Neanderthals and early Moderns an understanding of foraging strategies is critical (Marean & Kim, 1998). The initial success of hominids in exploiting open savannah environments may lie partly in the spatio-temporal mapping memory of ancestral tropical forest frugivores (Milton, 1981). After 2 Myr a cooler and drier, and more seasonal climate made fruit a less dependable source of food. There was therefore a shift to underground foods such as tubers, which are relatively abundant in the savannahs. Speth (1989, 1991) considered that there were physiological limits to total protein intake and that meat consumption was therefore kept at moderate levels by early hominids. Bunn & Ezzo (1993)

94

considered the importance of roots and tubers as efficient stores of nutrient and water with the added advantage of availability over most of the year and resistance to fire and drought. These would also have been easy to collect and so they may have been of some importance in the ancient hominid diet.

Nevertheless, where forests gave way to savannahs at the end of the Pliocene in East Africa, carbohydrates may have become the limiting nutrient in early hominid environments, requiring compensation through higher intakes of protein and fat (Bunn & Ezzo, 1993). Metabolic adaptation to long-term intakes of high levels of protein is experimentally demonstrable. In any case there is clear evidence of large-scale meat processing as from 2 Myr (Walker, 1981) which increased significantly with archaic *sapiens* forms, especially the Neanderthals (Foley, 1992). Foley (1989) relates the appearance of Moderns to increased foraging efficiency and the utilisation of animal resources. The use of meat appears to have evolved as a mechanism for enhancing flexibility for coping with periodic uncertainties in the food supply, given that in tropical savannah systems a range of mammals in a wide state of physical conditions would have guaranteed year-round availability, and provided critical nutrients and energy (Bunn & Ezzo, 1993). Muscle meat is, in particular, a valuable energy source and an important store of food that might sustain for considerable time periods a population accustomed to irregular feedings and unpredictable food resources (Bunn & Ezzo, 1993).

The response to spatio-temporal variability by hunter–gatherers is resolved by averaging out over time and space. Fat deposition, storage and other cultural buffers (e.g. food sharing, Kaplan & Hill, 1985) can do this. Trade can replace mobility as a way of averaging over spatial variation when increased competition requires greater productivity (Cashdan, 1992). A greater reliance on storage is associated with a decrease in mobility (Binford, 1980). Other authors consider that storage is compatible with mobile societies in the form of strategically placed caches in a seasonally revisited landscape (Stopp, 2002), a very likely tactic of the Moderns in the cold environments of the Eurasian Plains. Storage among hunter–gatherers buffers predictable seasonal variation and is therefore most common in highly seasonal environments. With an increase in storage there is a decrease in the number of residential moves (Kelly, 1983). Humans today are chararcterised by having among the highest levels of adipose tissue of all mammals (including Arctic species) and this may be a relic of having evolved in highly seasonal and unpredictable environments (Pond, 1978, 1999). Storing fat in adipose tissue permits humans to build up a considerable energy reserve and fat soluble vitamins (Bunn & Ezzo, 1993). These authors conclude that Plio-Pleistocene (and later) hominids who faced continuous uncertainty in their food supply had the problem of balancing essential nutrients and that this might only have been met by the presence of stored nutrients and energy in fat that could make up for dietary imbalances. Individuals carrying fat do not have to draw on muscle tissue to meet their energy needs so that individuals with an ability to store fat would have had a fitness advantage. One way in which fat reserves could have been accumulated was through gorging on meat, as occurs in contemporary San bushmen and Hadza (Bunn *et al.*, 1988; Hitchcock, 1989).

Humans are neither strict hunters nor scavengers. It is clear today that Neanderthals and Moderns were opportunistic and hunted, foraged or scavenged depending on circumstances (Marean & Kim, 1998). Research papers using stable isotope (Bocherens *et al.*, 1991, 1999; Fizet *et al.*, 1995; Richards *et al.*, 2001) and buccal microwear analyses (Lalueza *et al.*, 1996) have led the authors to conclude that Neanderthals consumed mammalian herbivore meat. The stable isotope data come from five specimens in three central European sites in Belgium, France and Croatia, all on or at the edge of the Eurasian Plain. They span a huge period of time. The buccal microwear data come from a few more sites, mainly in French but also from the Levant and the Gibraltar Devil's Tower child. Nevertheless, though suggestive, the data are too limited to generalise across the entire geographical range and the huge spans of time involved. The conclusion that the Gibraltar Devil's Tower individual was mainly carnivorous (Lalueza Fox & Pérez-Pérez, 1993) illustrates the difficulty. If we assume that this individual was not anomalous then, on the ecological evidence from the Gibraltar sites (Finlayson & Giles Pacheco, 2000; see below), we would have to accept that, while a large proportion of the diet would have been meat it would not have been exclusively so.

The problems associated with high protein intake, especially in skeletally-robust hominids, and the need to include fat and carbohydrate have been outlined by Cachel (1997). Additionally, high protein intake may have a negative effect on pregnant women which may explain reductions in protein intake by hunter–gatherers at certain times of the year (Speth, 1991). In the Mediterranean lands Neanderthals may have had difficulty in obtaining fats from mammalian herbivores that would have been leaner than their counterparts in the high latitude plains. This difficulty may have been alleviated by consumption of marine mammals – in Vanguard Cave, Gibraltar, during the last interglacial Neanderthals consumed monk seal *Monachus monachus* and probably dolphins. The options presented by the Mediterranean environments in terms of insect larvae, fruits, nuts, roots and tubers and marine molluscs in coastal sites, appear to have been exploited by Neanderthals, thus minimising the effects of total dependence on mammalian meat. The conclusion that Neanderthals consumed mammalian herbivore meat is undeniable, it has been known for a long time. What we cannot infer, however, is that that is all they consumed. Perhaps the individuals on the plains only ate such meat but, then again, there would not have been much else that could have been eaten and herbivores would then have been abundant. Plains dwellers would initially have had access to terrestrial mammalian fat and freshwater fish in localised areas. These Moderns were less robust than the Neanderthals and the problem of loss of calcium resulting from a high protein intake would have been reduced (Cachel, 1997). It is interesting to note that in Arctic hunter–gatherers, group size and sociality is constrained by the cost of acquiring adequate amounts of fat (Cachel, 1997). The ease of acquisition of fat by Moderns exploiting the herbivore biomass on the Eurasian Plain may be a contributory factor in the sociality and large group sizes of these people (Gamble, 1999; see below).

There is a view that links changes in food types consumed by humans and an increase in the diversity of food types taken from the middle Palaeolithic to human population pulses (Stiner *et al.*, 1999, 2000). According to this view humans would have initially selected 'slow' prey, that is prey in which capture time was minimised, and then moved to more mobile prey once the slower prey had been depleted (Stiner *et al.*, 1999, 2000). My view is that the only generalisation that we can make about diet is that humans have for a long time been able to eat a wide range of foods. People from at least the time of the common ancestor of Neanderthals and Moderns have been opportunistic omnivores capable of handling a wide range of foods, animal and almost certainly vegetal. I predict spatio-temporal differences at all scales in response to spatio-temporal resource heterogeneity. There is no theoretical reason or empirical evidence to propose that changes across time should be linear or unidirectional. If there is a case to be made for the diversification of the range of prey exploited and methods used by humans, then it is only after the Last Glacial Maximum (LGM) and especially towards the Pleistocene–Holocene boundary as large mammalian herbivores became regionally depleted (Holliday, 1998; Elston & Zeanah, 2002). The subsequent evolution of food production may be a development of this process (Chapter 8; Diamond, 2002)

Stiner *et al.* (1999) suggest that Palaeolithic human population growth depended on variations in small game – overexploitation depressed the populations of certain prey leading to hunting of less favourable types. Only four Italian and two Israeli sites were used in the analysis and the temporal scale of resolution of the faunal data did not match the finer-scale variability of the late Pleistocene climate (Chapter 6) so that it is not possible to conclude that communities were insensitive to climate. The claim is made on the basis of species composition comparisons and does not take abundance into account. Their analysis of prey composition through time is flawed. Inter-site data are lumped. The relative contributions of marine mollusc and vertebrate abundance are compared even though the methods of estimation differ – number of identified skeletal specimens are used, instead of minimum number of individuals, to estimate vertebrate prey but minimum number of individuals are estimated for marine molluscs and the two are readily compared. Claimed trends in prey size reduction are statistically insignificant with considerable overlaps. In any case the size differences may reflect inter-site differences. For example, all their estimates of humeral shaft diameter for the early periods (200–70 kyr) are from Hayonim Cave in Israel and it is impossible to determine whether later changes are due to temporal shifts or simply because other sites (which might always have had smaller size categories) were being sampled. Even more seriously, in the case of marine molluscs different species within the same genus (e.g. *Patella*), that are known to differ in size in the wild, are lumped in size comparisons and we are therefore left with the uncertainty of the extent to which the observed trends simply reflect different proportions of species in each sample. Possible biases due to inter-site and inter-species variations may therefore be superimposed on the claimed temporal patterns and these alternatives have been overlooked.

The availability of tortoises would have been reduced during Oxygen Isotope Stages (OIS) 4 to 2. Northern Italy is currently at the edge of the geographic range of *Testudo hermanni* and well outside that of *T. graeca* (that in any case may be a recent introduction); in Stiner *et al.*'s (1999) study, the tortoise disappeared faster in Italy than in Israel where they were not lost altogether. The reductions in sea level generated by cooling would have disconnected the Italian caves from the immediate coastal environment (Kuhn, 1995) and could have also reduced the marine mollusc contribution. Such environmental factors could also explain their observations. Relative abundance trends in other prey would result from tortoise and marine mollusc reduction and need not reflect real increases. Recent work in Gibraltar (Finlayson & Giles Pacheco, 2000) indicates that vertebrate community composition was similar throughout the late Pleistocene but climate altered vegetation and the local availability of species. Stiner *et al.*'s (1999, 2000) study cannot even be regarded as indicative because it extrapolates from the scale of a handful of local sites, some of which may not even be independent of each other, to a global scale. Whether or not there was ever a broad spectrum revolution (see Chapter 8) we certainly cannot infer it from these studies. Ultimately, climate seems to have been the key factor in the affairs of the Palaeolithic humans of the Mediterranean and further.

Clearly, in more varied regions Neanderthals were omnivores. It is more likely that the over-dependence on meat in marginal areas reflects the increasing stress to which these populations were subjected. The reality is that we have increasing evidence that Neanderthals across a huge time span and going as far back as the last interglacial at least were consuming marine resources including molluscs, seals and probably fish and cetaceans, just as other contemporary humans were doing in similar situations at the same time in South Africa (Deacon, 1989).

Deacon (1989) has argued that African Middle Stone Age (MSA) subsistence behaviour should be regarded as 'modern', there being no evident difference in subsistence ecology. Acheulian sites in South Africa are tied to valleys and water sources in the coastal platform. MSA/LSA (Late Stone Age) sites are found high up in the Cape Mountains as well as on the coast and there is frequent use of rock shelters. MSA populations ate meat and marine and molluscs (source of minerals) but there is no evidence of fishing or hunting of flying birds. We also know that in the right conditions, for example in central Africa, the harvesting of freshwater resources was happening in the MSA (Brooks *et al.*, 1995). Similarly, the Neanderthals occupying the topographically heterogeneous Mediterranean belt from Iberia in the west to, at least, Crimea in the east exploited a wide range of foods that included large mammals, small mammals and birds, tortoises, marine molluscs and probably even marine mammals, fish and plants (Stiner, 1994; Finlayson *et al.*, 2000a). Such a varied diet was probably a reflection of the micro-spatial and seasonal variability in resource availability in these areas and these would have also reduced the risks associated with overdependence on specific prey items. In Israel Moderns and Neanderthals hunted the same animals but Moderns differed from Neanderthals in having a more seasonally-specific hunting strategy (Liebermann & Shea, 1994).

Humans have therefore been consuming a broad spectrum of prey, when available in suitable environments, from at the very least the last Interglacial and probably much further back. The opportunistic humans would have optimised foraging tactics and these would have varied temporally and spatially, and at different scales, depending on resource availability. The degree to which Moderns and Neanderthals were specialised hunters is also likely to have been very flexible. Clearly, Moderns in the open plains of Eurasia probably specialised in particular types of herding prey at particular times of the year but Neanderthals on the edge of the plains were probably very similar (Gamble, 1986; Mellars, 1996). Moderns and Neanderthals in the heterogeneous mid-latitude belt would have varied from specialised to generalised hunting in accordance with the nature and dispersion of their prey. I would therefore predict a higher probability of specialisation on the open plains than in the more heterogeneous landscapes. In the latter case I would expect a mosaic of specialisation–generalisation, related to environmental and climatic features, that is independent of hominid taxon. Recent evidence from south-western France (Grayson & Delpech, 2002) that shows specialisation in particular resource taxa by Moderns and Neanderthals corroborates this view. It is hardly surprising that the authors should find no difference in the level of specialised hunting between the Mousterian and Aurignacian of this region. The two populations responded to similar terrain in a similar fashion, a situation not dissimilar to that in the Levant (see below). In any case we must be aware that the meagre data available to us lacks the resolution that some authors would like and it is not possible to substantiate global theories on this basis.

Gamble (1995) compiled a database of 588 sites in his north-central (NC), south-east (SE) and east Mediterranean (ME) regions. These regions coincide approximately with the Eurasian Plain (NC), the heterogeneous mid-latitude belt (ME) and an intermediate region (SE) between the two. Gamble (1995) provided data from archaeological sites and palaeontological sites, the latter with no human activity. Since the data recorded presence or absence of species in each site, density biases were avoided. I have re-analysed these data (Table 5.1) – I estimated species availability to humans from the palaeontological data. The data in Table 5.1 provide the following information: (a) availability – the frequency of each species in each region; (b) selectivity – the difference between presence in archaeological sites and the expected presence from the palaeontological data; and (c) relative differences in selectivity between regions within time periods allocated to Middle Palaeolithic, early Upper Palaeolithic and late Upper Palaeolithic. The data were too fragmentary for the late Upper Palaeolithic to be compared with the other two periods. The following patterns emerged from the data.

Predominantly plains species

These were mammoth, horse and reindeer. All three were actively selected by humans on the Plains (selected equates to hunted or scavenged in all cases). The availability of the three species in the mid-latitude belt (MLB) was low. Mammoths were selected as encountered but horse and reindeer were actively selected. Mammoth and reindeer were selected in the plains at a higher rate than in the MLB in the Middle and early Upper Palaeolithic. Horse was also selected at a higher rate in the Plains in the Middle Palaeolithic but the trend was reversed, although not in equal intensity, in the Upper Palaeolithic. This trend may reflect the southern range shift of the horse with the onset of the LGM. The dataset is incomplete for the giant deer, the elk and the saiga but we may tentatively place them as Plains species occurring at intermediate levels of availability, the first two species being actively selected and the saiga selected at the rate of encounter. The availability of giant deer in the MLB was low but they were actively selected. Elk and saiga appear to have been largely absent. There is a suggestion of a trend towards higher rate of exploitation of giant deer in the MLB in the Upper Palaeolithic, possibly reflecting a similar range change response to that for the horse.

Table 5.1. *Analysis of mammalian herbivore consumption by humans in Europe. The first three columns record the availability (AV) of each species for all regions and time periods: +, occur statistically significantly greater than expected; =, no statistical difference from expectation at random; −, occur statistically significantly below expected. pl, Plains; int, intermediate region between plains and heterogeneous belt; het, heterogeneous belt. The next three columns record the selectivity (Sel) by humans for all regions and time periods. The remaining columns record the observations by time periods: MP, Middle Palaeolithic; EUP, Early Upper Palaeolithic; LUP, Late Upper Palaeolithic. Statistically significant positive relationships are in dark grey; insignificant relationships are in light grey; statistically significant negative relationships are in white. See text for interpretation*

Species	Av			Sel			MP			EUP			LUP		
	pl	int	het	pl	int	het	pl	int	het	pl	int	het	pl	int	het
Mammoth	+	=	−	+	=	=	+	−	−	+	−	−			
Rhinoceros	=	=	=	+	+	=	+	=	=	+	=	=			
Bos	=	=		+	=	+	=	=	=	−	=	=			
Megaceros	+			+	=	+	=	=				+			
Alces	+	=	−	+	+		+	=	−						
Horse	+	=	−	+	+	+	+	=	−	=	=	=	+	=	−
Red deer	−	=	+	+	=	+	=	=	=	=	=	=		=	=
Reindeer	+	−	−	+	+	+	+	=	−	+	=	−	=	=	=
Sus	=	=		=	−	+	=	=	=	=	=	=	=	=	
Ibex	−	+	+	+	=	+	−	=	+		=	=			
Chamois					+										
Saiga	=														
Roe deer		=	=	=	−	=	=	=	=						
Selected				9	4	6	5	0	1	3	0	1	1	0	0
Encountered				4	5	4	4	7	5	4	6	4	1	2	1
Avoided				0	2	0	1	4	5	1	2	3	0	0	1

From: Gamble (1995).

Predominantly heterogeneous landscape species

These were red deer and ibex. Ibex were actively selected but red deer were selected as encountered. The availability of the two on the plains was low and both were actively selected. Ibex were selected at a higher rate in the MLB than on the plains in the Middle Palaeolithic but there was no difference in the early Upper Palaeolithic suggesting a greater specialisation in ibex hunting in the Middle Palaeolithic in the MLB. There was no difference in the case of the red deer, between regions or periods. The dataset is incomplete for the wild boar and chamois but we may tentatively place them as MLB species occurring at intermediate and high levels of availability respectively and both actively selected. Both probably occurred at low availability in the plains, wild boar being selected at the rate of encounter and chamois being actively selected.

Intermediate species

These were rhinoceros and aurochs. Gamble (1995) does not differentiate between rhinoceros species. If he had, differences between Plains species and those of more vegetated habitats may have emerged. Rhinoceros and aurochs occurred at intermediate levels of availability on the plains and the MLB. Rhinoceros were actively selected on the Plains and selected as encountered in the MLB. The pattern was reversed for aurochs. The dataset is incomplete for the roe deer but we may tentatively place it as an intermediate species occurring at intermediate levels of availability on the plains and the MLB and selected as encountered in the two regions.

Humans therefore appear to follow particular prey selection strategies that we may summarise as follows:

(1)   There appears to be a greater specialisation in the plains where most species are actively selected. In the MLB many more species appear to be selected as they are encountered. This difference may explain claims that Neanderthals hunted prey as it was encountered and Moderns by planned searching of particular prey species. We can see how the predominance of Neanderthal sites in the MLB and Moderns sites on the plains can lead to this apparent pattern.

(2)   A number of prey species are actively selected in situations in which they occur at high density. On the plains we have the three main herding species: mammoth, horse and reindeer. On the MLB we have the two rocky habitat herding species: ibex and chamois. There are no cases of prey that occupy intermediate or closed vegetation in this category.

(3) A number of prey species are actively selected even though they occur at low levels of availability. On the plains these are typical MLB species: ibex, chamois and red deer. On the MLB they are typical Plains species: horse, reindeer and giant deer. Four are herding species: ibex, chamois, horse and reindeer, that were also entered in (2). Red deer is also a herding species that would be accessible in open country as would be the case in the plains. Giant deer may be a similar case.

(4) There are several species that are actively selected but occur at intermediate levels of availability. The reasons for their selection may lie in a combination of the factors described in (2) and (3). On the plains these species are giant deer, elk and rhinoceros. On the MLB they are aurochs and wild boar.

(5) The species that are selected at the rate of encounter are species that are either: (a) rare in marginal geographical areas – mammoth in MLB and wild boar on the plains; or (b) species that are dispersed in vegetation and rarely venture into open vegetation – roe deer everywhere, aurochs on the plains and rhinoceros and red deer on the MLB.

(6) The saiga appears anomalous. It is an open plains species that can aggregate and would have occurred at times at intermediate or even high levels of availability. Two reasons may explain the anomaly. The species was sporadic in Europe or its small size reduced its appeal to human hunters.

These results support the view that mammalian herbivore exploitation by Pleistocene humans was related to ecology and not to the human type. There are very few obvious shifts in prey exploitation between the Middle and early Upper Palaeolithic and when they occur, as with the horse, they appear related to shifting ecological boundaries.

While there may be a case for using 'overkill' hypotheses in the case of colonising human populations, such as those arriving in the plains of Eurasia at the end of the Pleistocene (and I am not totally convinced), such an argument would seem to have little value when examining well-established populations of hunter–gatherers such as the Mediterranean Neanderthals. The most probable relationship between Neanderthals and their resources would have been one of density-dependent population regulation and not over-exploitation. In the absence of fine-grained data showing the contrary this must remain the most ecologically plausible and parsimonious explanation. Furthermore, the often rapid climatic oscillations of the Pleistocene in Europe would have generated continuous range and density shifts in many species that were consumed by Neanderthals. In such situations of instability abiotic factors would have been the key to continuously alter prey densities.

I therefore conclude that there were significant dietary differences between peoples (modern or archaic) inhabiting the northern plains (largely mammal-meat consumers) and those in the heterogenous landscapes to the south (where they had broader diets) (Table 5.2). Those in the tropics would have had, as they do today, the greatest available range of foods. It would have been the plains dwellers that evolved the most sophisticated behavioural and physiological risk reduction tactics.

Habitat, landscape and geographical range

The only quantitative study of Neanderthal habitat that looked at vegetation structure as well as species composition was the one that examined the Gibraltar Neanderthals (Finlayson & Giles, 2000). In that study I demonstrated that Neanderthals in Gibraltar hunted in what I described as a Mediterranean wooded savannah, that is a fairly open vegetation with a mix of shrub and light tree cover (Figure 5.1). In other words, Neanderthals were exploiting situations that were of an intermediate structural nature, that is neither dense forest nor fully open plains. Such ecotones or areas of high habitat heterogeneity are expected to be high in mammal species richness (Kerr & Packer, 1997). I also showed how a change towards dense montane forest at the end of OIS 3 corresponded to the disappearance of the Neanderthals from the area (Figure 5.2). Neanderthals also exploited other habitats, specifically cliffs and similar rocky areas, estuaries and coastal habitats. The evidence from other regions shows that, as in Gibraltar, Neanderthals exploited intermediate habitats between closed forest and open plains (Soffer, 1994; Mellars, 1996). These habitats would have suited them well as they would have had a rich grass layer that would have been attractive to grazers (Finlayson & Giles, 2000). There would have been some cover for prey to be stalked and the cover would not have been too dense to restrict hunting and herbivore activity.

Another link between Neanderthals and habitats comes through fresh water. The Gibraltar Neanderthals would have had ample supplies of freshwater close by (Finlayson & Giles, 2000). In the Perigord, south-west France, the distribution of Neanderthals is close to rivers (Mellars, 1996) and the association between Neanderthals and other contemporary humans with lacustrine and other freshwater habitats seems to be a widespread and trans-continental phenomenon (Nicholas, 1998). The association would seem to have a dual advantage: the availability of drinking water and the attraction such habitats have for other animals and therefore as a source of prey.

The distribution of Moderns in the early stages in Eurasia is associated with open plains habitats (Soffer, 1985; Finlayson, 1999; Finlayson *et al.*, 2000a).

Table 5.2. (a) Summary of predicted utilisation of food resources by late Pleistocene humans on the Eurasian Plain and the Eurasian mid-latitude belt. Main resources are in dark grey cells. Important resources are in pale grey cells. (b) Summary of predicted habitat use by Late Pleistocene humans

(a)

| | Mammalian herbivores | | | | | | Birds | Tortoise | Fish | Marine molluscs | Fruit |
	>1000 kg	1000–500 kg	500–100 kg	<100 kg	Small mammals	Marine mammals					
Eurasian Plain	+++	+++	+++	+	+	–	++	–	++	–	+
Mid-latitude Belt	+	++	++	++	++	++	+++	+++	++	+++	++

(b)

	Closed	Intermediate	Open	Rocky	Wetland	Coast
Eurasian Plain	–	+	+++	+	++	+
Mid-latitude Belt	+++	+++	+	+++	++	+++

This immediately suggests a difference in habitat use between Moderns and Neanderthals. There is a range of habitats utilised by Moderns that includes the types used by Neanderthals and all we can conclude, on present evidence, is that Moderns included open plains as habitats that could be exploited much more intensely and frequently than did Neanderthals.

The preference for intermediate structural habitats by Neanderthals is also detectable at the landscape level. At this level, a number of studies from such diverse geographical regions as Iberia (Finlayson & Giles, 2000), south-west France (Mellars, 1996), the Middle East (Shea, 1998) and the edge of the Russian Plain (Soffer, 1994) show beyond doubt that Neanderthals occupied landscapes that were ecotonally rich – that is landscapes that included a diversity of habitats over a small area. Topographically heterogeneous regions are especially diverse. Other important ecotonal landscapes that appear to have been repeatedly used by Neanderthals are wetlands, coastal landscapes, lake mosaics and linear riverine stretches. The advantage of such areas is that they

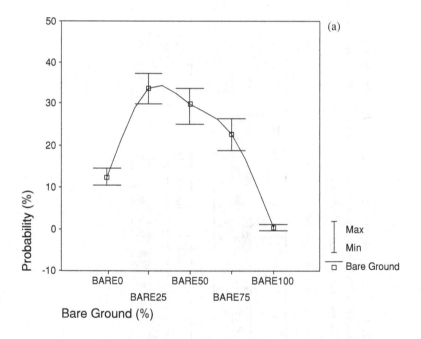

Figure 5.1. Predicted patterns of vegetation structure in the Neanderthal Oxygen Isotope Stage 3 site of Gibraltar (after Finlayson & Giles, 2000). (a) Distribution of bare ground; (b) distribution of tree heights; (c) tree density (trees/ha.); (d) distribution of shrub heights; (e) distribution of grass heights; (f) cover of stone pine *Pinus pinea*; (g) cover of juniper *Juniperus phoenicea*; (h) distribution of trees by trunk circumference.

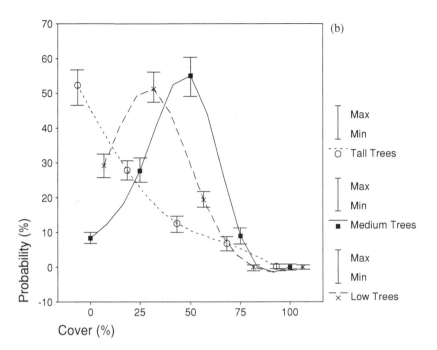

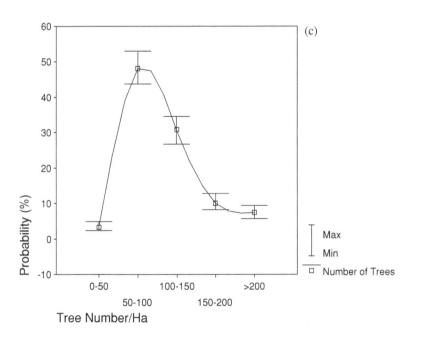

Figure 5.1. (*cont.*)

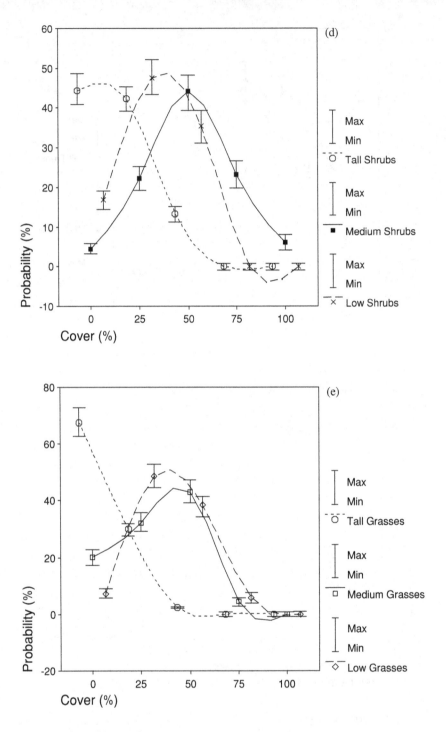

Figure 5.1. (*cont.*)

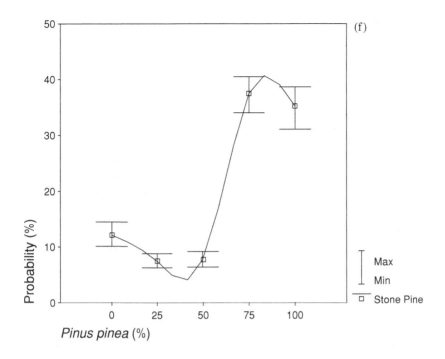

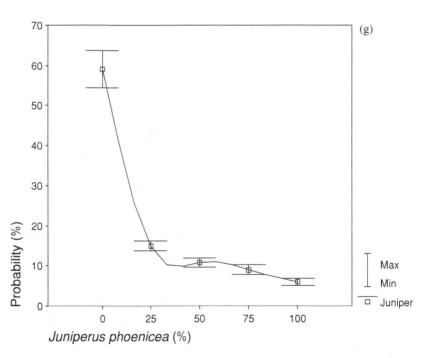

Figure 5.1. (*cont.*)

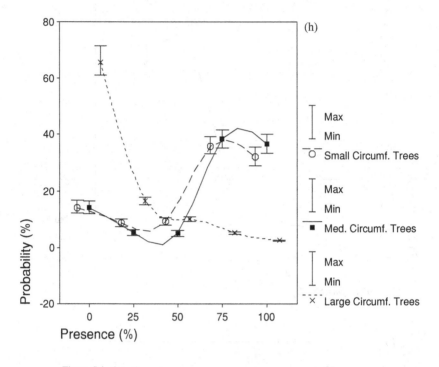

Figure 5.1. (*cont.*)

contain a rich diversity of potential prey over short distances. These landscapes, by providing a range of options, are also more buffered than more homogeneous landscapes against environmental instability.

Because open plains habitats stretch across great areas over topographically homogeneous land, Moderns utilising open plains de facto occupied homogeneous landscapes. There are several requirements for successfully exploiting homogeneous open landscapes: (a) because a small range of highly mobile prey, that may occur in localised areas at high density, are available, predators will need to have behavioural tactics that allow for such mobility; (b) because of the lack of cover predators will require specific and specialised hunting techniques; (c) because of the wide ranging nature of this behaviour pattern, and the need for sources of water especially if high levels of meat are consumed, predators will not operate well if such landscapes are in arid regions; and (d) complex social systems are necessary to increase environmental resistance, as we shall see later. The Eurasian Plain may have been exceptional if the humans exploiting it were able to melt the readily available snow and ice.

In Chapter 4 I discussed the evidence for climatic adaptation in Neanderthals and Modern Humans. Here I relate the evidence to geographic distribution

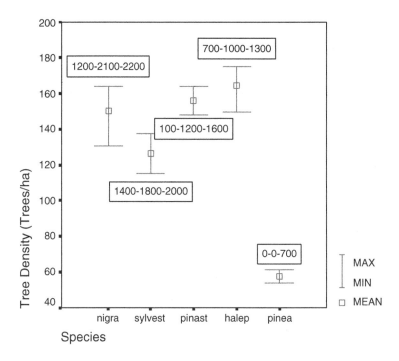

Figure 5.2. Present-day density of pine (*Pinus*) trees in southern Iberia. Boxes indicate altitude range in metres. nigra, *P. nigra*; sylvest, *P. sylvestris*; pinast, *P. pinaster*, halep, *P. halepensis*; pinea, *P. pinea*. Contrast present distribution of *P. pinea* (low altitude) with *P. nigra* (high altitude). During late Oxygen Isotope Stage 3, *P. nigra* was the dominant pine in Gorham's Cave, Gibraltar, currently at sea level.

patterns. It is widely accepted that Neanderthals were cold-adapted. I have no doubt that populations of humans (Neanderthals or Moderns) living in cool climates would have benefited from adaptations that minimised heat loss. The question is to what extent were such adaptations significant in defining geographical range limits? The evidence suggests that they were unimportant. Late Neanderthal sites are all in Mediterranean or sub-Mediterranean bioclimatic zones and cool temperate sites are associated with warmer periods (Figure 5.3). In Chapter 4, I showed that Neanderthals would have evacuated geographical areas long before the arctic conditions characteristic of the strong glacial pulses would have reached them. Instead, we have the Moderns occupying the plains of Eurasia once the Neanderthal range has receded towards the south. The Moderns too were hit by the glacial cold, but significantly later and when it reached greater intensity towards the LGM (once Neanderthals were extinct). The range of the Moderns only then extended south into the Mediterranean peninsulas of Iberia, Italy and the Balkans, as suitable habitats closed in the north and opened up in

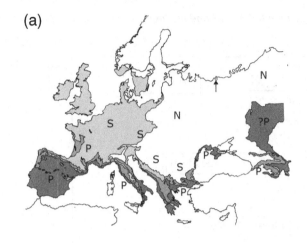

(a)

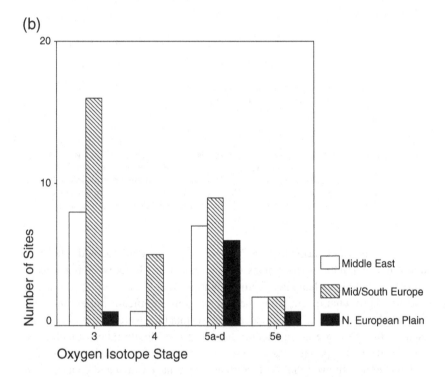

(b)

Figure 5.3. (a) Occupation of Europe by Neanderthals. (b) The histogram (based on data in Stringer & Gamble, 1993) illustrates the sporadic occupation of the North European Plain, only during warm intervals. P, Permanent occupation; S, semi-permanent occupation; N, never occupied. Bioclimates after Rivas-Martínez (1996). Dark grey, Mediterranean; intermediate grey, sub-Mediterranean; pale grey, temperate oceanic; white, temperate continental (below arrowed line) and boreal (above arrowed line).

the south. This raises the question of the extent to which the supposed tropical morphology of the Moderns meant anything to them. They clearly survived the cold of the north until they were forced south because of habitat and resource loss. Clearly behavioural and physiological adaptations including the use of fire, shelters, clothing, the inclusion of fat in the diet, food caching and a complex social system were overriding factors for Moderns just as they are for Eskimos today. Eskimos today do not differ from central Europeans in their ability to retain heat. They simply wear suitable clothing and consume large amounts of animal fat that permits a higher rate of non-shivering thermogenesis (Gisolfi & Mora, 2000). This begs the question of the extent to which comparisons with modern populations are valid. The crucial point is that it was food and habitat, not temperature, that limited the geographical distribution of Eurasian humans.

Home range, group size and related features

The spatial arrangement of resources, habitat and barriers affects the location, movement patterns, foraging dynamics and persistence of organisms (Karieva, 1990; Danielson, 1991; Pulliam *et al*, 1992; Turner *et al*, 1995). In humans, the way they move in the landscape exerts strong influences on culture and society (Kelly, 1992). In the case of the polar bear, a species that has evolved into a High Arctic carnivore from an omnivorous ancestor, movements are constrained by landscape pattern (Ferguson *et al.*, 1998). Home-range size in polar bears is related to the ratio of land to sea and to the seasonal ice cover variation. Polar bears adjust the size of home range according to the amount of annual and seasonal variation within the centre of the home range. Bears experiencing unpredictable seasonal and annual ice respond by increasing their home-range. The effect would be a reduction in the variation in seasonal and annual ice that they experienced. These animals therefore make trade-offs between alternate space-use strategies. Large home ranges occur when there is variable ice cover that is associated with more, but also more unpredictably distributed, seals (Ferguson *et al.*, 1999). In the case of baboons, predation, group size, home and day range all increase with aridity (Dunbar, 1984). The critical variables in their socio-ecology are food quality, food patchiness and predictability. The proximate mechanism behind the relationship between foraging time and ecological variables in baboons is mediated through the impact of climate (temperature and seasonality) on vegetation structure and food availability (Hill & Dunbar, 2002). In this book I emphasise the critical relationship between human behaviour and vegetation structure and food availability, the latter variables being strictly influenced by climate. Such relationships appear to explain, more widely, much of the variation in mammalian behaviour across a range of spatial scales (Hill & Dunbar, 2002).

Foley (1992) proposed that the occupation of open habitats by the earliest hominids meant that they were exploiting environments with greater uncertainty, seasonality and patchiness, and that the expected adaptive response would have been an increase in day and home ranges. A related outcome would have been a larger annual range. The earliest hominids were found associated with indicators of habitats with large patches of grassland and bushland. Such environments promoted larger group sizes and extended foraging ranges in primates (Foley & Lee, 1989).

The early members of the genus *Homo* continued the open habitat colonisation process and increased efficiency with the use of tools, an increase in the consumption of animal resources, increasing further day and home ranges, and also group size (Foley & Lee, 1989; Foley, 1992). Male co-operation in food acquisition became advantageous and the defence of females increasingly difficult. Co-operative defence of groups of females minimised loss. This enhanced male kin relationships. The frequency, stability and intensity of male–female associations increased as a result of higher maternal reproductive costs due to increased encephalisation, slower infant growth rates, delayed maturity/independence, infants at additional risk of non-nutritional mortality due to predation, intergroup encounters and infanticide (Foley & Lee, 1989). Such adaptability permitted *H. ergaster/erectus* to colonise more temperate and seasonal climates (Foley, 1992). Even so the distances involved in movement, as measured by raw material transport, rarely exceeded 60 km in the Lower Palaeolithic (Feblot-Augustins, 1990). By OIS 6–5 distances of 100 km were becoming increasingly frequent in Middle Palaeolithic Europe and it is in the Upper Palaeolithic that we observe the greatest increases, reflected by quantities of raw materials imported from distant sources and in terms of stages of manufacture in which the lithics were transported over these greater distances (Feblot-Augustins, 1993). I shall interpret the significance of these increases below.

The ecomorphological characteristics of Moderns (Chapter 4) were a further development of this continuing process and it enabled them to increase the use of high quality resources, increase further the size of home ranges, group size and structure. This was achieved by improving the ability to ensure hunting and foraging success, thus turning unpredictability into predictability. Greater longevity allowed for extension of information across generations and therefore the recognition of lineage through space, thus structuring relationships between groups over the long term. The final stage of the process, according to Foley (1992), was the evolution of food production in the Neolithic.

I agree with Foley (1992) that these trends continued right up to the beginning of the Holocene. Thus, the evolving Moderns in Africa during Stage 6 developed new skills, leading to better use of raw material, and new ecological attributes like larger home ranges, day ranges, group size and dietary selectivity

(Foley, 1989; Foley & Lahr, 1997) This, in my view, occurred when this population was expanding and when the pre-Neanderthals would have had restricted ranges in Europe because of the prevalent cold conditions. Foley (1989) has proposed some ecological features of Moderns and these are characteristic of patchy environments with high quality and predictable food: (a) a very large home range; (b) a very large day range; (c) a large group size; (d) a high degree of sub-structure within groups; and (e) dietary selectivity towards meat, which occurs in large packages. In my view mobility strategies and home ranges are principally dependent on the availability, dispersion and accessibility of food resources. Availability of raw materials for tool manufacture, though undoubtedly important (Feblot-Augustins, 1993), would have played a subsidiary role except in cases of food resource superabundance and lithic raw material localisation or scarcity.

Studies of contemporary hunter–gatherers lend support. Harpending & Davis (1977) argued that ranges will be smaller where resources are in phase since people can remain near sites of resource concentration. Increasing range in areas with low patchiness does not increase diversity. Therefore larger ranges will occur with patchy resources that are out of phase. Kelly (1983) showed that hunter–gatherers in colder climates move a greater distance with each residential move because resource patches are out of phase at low frequencies. The Hadza and !Kung San live in arid savannahs. The !Kung live in areas where important resources (water and plants) are patchy and out of phase. The Hadza resources are more evenly distributed. The !Kung San have larger foraging ranges and !Kung camps are located near water sources. Food resources may be located many kilometres away. The Hadza can gather near the residential camp (Cashdan, 1992). There is a considerable body of evidence linking Pleistocene humans with wetlands (Nicholas, 1998). I have suggested elsewhere (Finlayson *et al.*, 2000a) that wetlands, as sources of concentrations of biomass especially in seasonal environments, would have been highly significant in the evolution of humans within tropical and Mediterranean zones and perhaps also in specific temperate situations.

Harpending & Davis (1977) have argued that if variance in total calories between ranges is large, people should be clumped in large groups at the rich locations. If variance is small, people should be more uniformly distributed and in small groups. The former would be the case for Moderns and the latter for Neanderthals. They also predict that: (a) if resources are distributed independently of each other, calorie variance between ranges will be low and occupations ought to be uniform; (b) if resources are correlated and in phase then variation from place to place will be high and the settlement pattern will be aggregated; and (c) if resources are out of phase then calorie variance between ranges will be small leading to uniform occupation of large ranges.

According to Binford (1980), when resources are patchy on a scale larger than the daily foraging range typical of boreal forest and arctic regions, the response of human hunter–gatherers is to locate the residential camp at one resource and send out small specialised task groups for other resources. These travel long distances and remain at other resource locales for several days or weeks (logistic mobility). Where resources are more evenly distributed a variety of resources is available near camp and, therefore, whole residential group moves as needed. Moves are shorter but more frequent (residential mobility).

A study of behavioural differences between Neanderthals and Moderns, based on a small sample of sites in Israel, indicated that Modern Humans were associated with seasonal patterns of resource (in this case gazelles *Gazella gazella*) exploitation whereas Neanderthals apparently exploited the same resource all year (Lieberman & Shea, 1994). This led Lieberman & Shea (1994) to interpret these data as indicative of Moderns having a circulating mobility pattern and Neanderthals a radiating mobility pattern. Circulating patterns require seasonal movements of groups between a series of temporary residential camps while radiating ones require movement from a more permanent (year-round) residential base camp to less permanent (seasonal) logistic camps situated close to important resources. They attributed the differences between Moderns and Neanderthals to morphological differences (Trinkaus, 1992) that reflected their different mobility pattern that, in Neanderthals, required higher daily expenditures of effort, and a higher level of hunting–gathering activity (Lieberman & Shea, 1994; Lieberman, 1998). According to these authors Neanderthals predominantly practised a radiating strategy, which required them to hunt more frequently than Moderns, and Moderns practised a circular strategy. They also accept that the two strategies need not have been mutually exclusive. The problem that I see with these interpretations relates to scale. As we have seen Moderns tended to be related with more open, homogeneous, landscapes than Neanderthals and they required larger annual territories. They probably practised a circular strategy, as proposed by Lieberman & Shea (1994), over the annual cycle and over a large territory. However, at each seasonal stop, they would have adopted a strategy closer to the radiating example with specialised groups performing different tasks, i.e. showing a high level of division of labour. The Upper Palaeolithic people of northern Aquitaine in France, for example, occupied winter settlements that were dedicated to the massive butchering of reindeer with alternative and functionally different sites in the summer (Demars, 1998). The utlility of this behaviour is seen in the Upper Palaeolithic exploitation of the Siberian subarctic or the central European Plain. These colonisers organised specialised hunting from small camps that had links to larger base camps, they had semi-subterranean dwellings, storage pits and art, they economised on raw materials and transported some over long distances

(Montet-White, 1994; Goebel, 1999). In other words they had the social, cultural and ecological package that characterised the Moderns of the open plains of Eurasia.

The key behavioural difference with the Neanderthals is that the Neanderthals operated at a smaller spatial scale (Feblot-Augustins, 1993) – they had smaller annual territories, probably reflecting the more diverse resource-base available in heterogeneous landscapes. Occupation of smaller annual territories would have permitted small-scale movement patterns of either a circulating or radiating type. Either way the effect would be to produce a picture of year-round exploitation. Thus the Levantine evidence is in keeping with the model proposed in this book. Working at a larger scale would have put premium on a group's ability to monitor the availability of distant resources, as indicated by Lieberman & Shea (1994). Were the two tactics interchangeable? In my opinion they were not for one simple reason – the difference in morphology. The heavy, robust, morphology of the Neanderthal would have precluded efficient exploitation of large territories. This may explain why robust archaic hominids never ventured into the steppe environments of the Eurasian Plain. Where they were close to it or on its margins, as in eastern Europe, they appear to have responded, as we would expect, by increasing their changing mobility strategy (Feblot-Augustins, 1993). Home range in mammals is generally predicted to increase with habitat fragmentation and the home ranges of large species will be especially sensitive (Haskell *et al.*, 2002). In orangutans, home range significantly increases in response to coarse-grained patchiness (Singleton & van Schaik, 2001). We would also expect increasing home-range size in Neanderthals in 'edge' regions (bordering the MLB and the plains), such as south-west France, in response to habitat fragmentation during periods of expansion of open vegetation. A more gracile morphology would have been far more efficient over larger areas. It may have sacrificed the close-quarter hunting strategy but this may have been compensated for by technology. Technological change would have made it possible for Moderns to be flexible in their use of large territory–open environment vs small territory–heterogeneous environment strategies.

So it is the gracile Modern Human strategy that we observe for the first time on the Russian Plain (Soffer, 1985). These people were entering unoccupied territory but one that was rich in mammalian herbivores. Their way of coping with the risks of colonising new territories may have a parallel in the subsequent colonisation of North America that I regard as part of the same process of expansion into the mammal rich environments of Eurasia and North America. The most suitable lifestyle in such new territories would have been one that placed primary reliance on faunal rather than plant resources, as such resources would be easier to locate. In addition, mammals would be available all year and would be widely dispersed. Different species could be dealt with by similar techniques. The process of adaptation to the plains continued throughout the late Pleistocene and populations occupied distant regions. The late Upper Palae-olithic hunters of north-east Asia were much more mobile than earlier Upper Palaeolithic peoples of the region and they concentrated their subsistence on a few key mammal species (Goebel *et al.*, 2000). In situations of fluctuating mammalian populations (due, for example, to climatic stresses) even the ex-ploitation of large home ranges would have been insufficient at times and a higher-scale tactic, that of frequently changing home range so that the home range exceeded the annual range, appears to have been the solution in the case of Upper Palaeolithic North American palaeo-indians (Kelly & Todd, 1988; Kelly, 1992). We can see how such a system could have equally applied to the colonisers of the Eurasian Plain. If so it would be further evidence of the degree to which risk was resolved by increasing the scale of operation.

Mobility is scale-dependent. To avoid confusion, I shall clarify my definition of mobility in Moderns and Neanderthals. Moderns, exploiting open environ-ments, had large annual (home) ranges. At that scale they were highly mobile. But their movement across this large home range was marked by periods of relative sedentism with base camps close to rich resources. At this lower scale, Moderns were more sedentary than Neanderthals. This needs to be qualified further. Division of labour in large groups, as would be expected in peoples exploiting large home ranges, would have meant that some individuals would have been sedentary close to the base camp while others would have ranged more widely in search of food resources. In the case of the Neanderthals mobil-ity would have been high at this lower scale and is likely to have been similar among all individuals. The precocity of Neanderthal young (Dean *et al.*, 1986; Trinkaus, 1986) comes as no surprise. At the larger scale, however, we may regard Neanderthals as more sedentary than Moderns as they covered a smaller geographical area during the course of the annual cycle. An understanding of the level of mobility of these populations is critical given that changes in mo-bility patterns are known to trigger dramatic changes in food storage, trade, territoriality, social and gender inequality, division of labour, subsistence and demography (Kelly, 1992).

The differences in spatial dispersion and mobility of hominids in heteroge-neous landscapes (Moderns and Neanderthals) and homogeneous landscapes (mainly Moderns) reflect wider patterns. According to Geist (1977) oppor-tunism in animals may be of two kinds. One is of a highly mobile nature that involves the exploitation of small food patches, as these become available, and subsequent movement on. This small-scale opportunism, that I equate with Neanderthals, ties individuals much less during reproduction and favours a small number of large, well-developed, highly mobile young that can follow the parent at an early stage. This is exactly what we observe in Neanderthals.

The second type also involves a mobile form but only until a large food resource is found, adequate for maintenance and reproduction. Since reproduction can be maximised there is selection for a maximum number of young and these disperse when they reach near-adult size. The small-scale opportunists, like the Neanderthals, are at the K-end of the selective strategy whilst the large-scale opportunists, such as the Moderns, are at the r-end of the range.

Technology

To understand technology and technological change we must relate it to ecological function (Kuhn, 1995). Foley (1989), for example, has proposed that the development of technology that allows predation from a distance, e.g. the use of projectiles, will provide the hunters with a major advantage. He tries to relate this to the very early appearance of blade technology in Africa. Prey can be unpredictable in distribution and success of capture, particularly in open environments where predators have the added disadvantage of not being able to get as close to prey as in more vegetated areas. Mechanisms that reduce this unpredictability will be favoured. Fitzhugh (2001) has recently provided a theoretical framework, using mechanistic principles of Darwinian evolution, that helps in our understanding of the observed technological changes in the Palaeolithic. Fitzhugh (2001) proposes a model of technological change that simply requires that individuals have the cognitive flexibility to assess their risk sensitivity and modify technological behaviour appropriately. Conditions of greater risk sensitivity should generate more inventive behaviour, especially at times of economic crisis. Since there is no guarantee of success for a novelty, the chances of replication of a novel behaviour will increase as inventive behaviour increases in response to greater risk sensitivity. As population density increases so we would expect that there would be more people available to generate successful novelties and that these should spread more rapidly due to greater facility of cultural transmission. The rate of technological change would therefore increase. This model predicts the changes observed in the Palaeolithic, including the slow early rate of technological innovation and its increasing rate of turnover that can be related to the increasingly unstable conditions (therefore increased risk sensitivity) of the last glacial cycle (Chapter 6). The increase in technological turnover in the Upper Palaeolithic additionally reflects the changing social circumstances described in this chapter as well as the increasing population density towards the end of the Upper Palaeolithic.

Diverse technological techniques were available to Moderns and archaic populations derived from a common *sapiens* ancestor since at least 150 kyr as proposed by Foley & Lahr (1997) and probably much earlier. The emergence of modern-type behaviour in different places at different times and its subsequent disappearance appears to be triggered by changing environmental conditions. Many features considered to typify the Upper Palaeolithic are present very early in Africa (McBrearty & Brooks, 2000). The presence of a well-developed bone industry, dated at around 90 kyr, in archaeological sites in Democratic Republic of Congo (Brooks *et al.*, 1995; Yellen *et al.*, 1995) brought this issue to the fore. Henshilwood & Sealy (1997) subsequently reported the discovery of two MSA bone points at the Blombos Cave in South Africa. McBrearty & Brooks (2000) have since argued that many of the components of the Upper Palaeolithic 'revolution' are indeed found in the African MSA tens of thousands of years before their appearance in Europe, suggesting the gradual assembling of the package of Modern Human behaviours in Africa with a subsequent export to other parts of the Old World. They claim to identify evidence of blade and microlithic technology, bone tools, increased geographic range, specialised hunting, use of aquatic resources, long distance trade, systematic processing and use of pigment, art and decoration. Most recently, Barham (2002a) has claimed the early presence of a backed tool technology, usually associated with late Pleistocene humans, in the middle Pleistocene of south central Africa and has placed its origins at around 300 kyr. The parallel distribution of pre-Upper Palaeolithic technology in both hemispheres would appear to lend support to the view of an early presence of modern behaviour (Ronen, 1992). The Amudian in Israel may date back to 150 kyr and Howieson's Poort in South Africa to 75–45 kyr. Some authors have reflected that tools, such as blades, are simply more sophisticated and frequent in the Upper Palaeolithic after 50 kyr (Bar-Yosef & Kuhn, 1999; Ambrose, 1998; 2001). This suggests that the changes between the Middle and Upper Palaeolithic are quantitative rather than qualitative, to do more with changing environmental circumstances and cumulative effects related to the spread of traditions than with changes in cognition. The presence of 400 kyr-old wooden spears in Schoningen, Germany (Thieme, 1997), opens up the question of the reliability of observations largely based on the study of lithics.

Moderns could therefore adapt using either modern or archaic patterns of behaviour. The differences between Middle and Upper Palaeolithic blade technologies appear to be simply related to economy of extraction (Ronen, 1992). Such technological adaptations also occur across different human forms occupying the same region and using the same Mode of Technology. Thus the differences between the Mousterian of Moderns and Neanderthals in Israel would appear to relate to different regional bioclimatic adaptations (Marks, 1992) and may well relate to the nature of the prey and environments being exploited (Shea, 1998).

We should not lose sight of the fact that lithic raw material availability, quantity, size and shape of nodules and flint texture, affect flaking techniques and artifact morphology and can therefore exert an influence on the observed assemblage patterns. In the Périgord Region of south-western France, for example, enclosed sites are characterised by a range of Mousterian assemblages with carefully reduced nodules and intensively trimmed implements, especially side scrapers. In nearby open-air sites, with greater access to abundant raw materials, assemblages are homogeneous and dominated by Mousterian of Acheulian tradition, with many hand axes and fewer retouched flake tools (Rolland & Dibble, 1990).

Van Peer (1998) has described the situation in north-east Africa. Some of the complex behavioural features studied in the Nile (raw material procurement, settlement systems, etc.) were present from the early Middle Palaeolithic. Towards the end of the Middle Palaeolithic technological experiments with the Levallois system were conducted in Lower Nile Valley. These were designed to produce blades and enhance core productivity around 40 kyr. In north-east Africa complexity was attained in different groups which have been equated to Moderns and archaics. The Aterian is considered to be an adaptation to a dry climate and may have been a solution that allowed Nubian groups to survive deteriorating climate (Van Peer, 1998). Modern behaviour was therefore not limited to Moderns alone in north-east Africa.

Changes in technology with changing environment have also been documented for Moderns and Neanderthals in the Middle Palaeolithic Levant (Shea, 1998) and for Moderns in Upper Palaeolithic Europe (Blades, 1999). Shea (1998) attributed differences in the proportion of Levallois points in assemblages to differences in intercept and encounter hunting associated with steppe and Mediterranean woodland habitats respectively. Shea (1998) attempted to relate these differences to Neanderthals and Moderns but his sample was too small to be meaningful. He, importantly, recognised the degree of variability, and hence flexibility and adaptability, of Neanderthals and Moderns in response to the exploitation of the two kinds of habitats. In reality Shea (1998) was examining a habitat mosaic at landscape level that would have included significant ecotones where the alternative strategies would have met. Blades (1999) studied the Aurignacian in the Périgord (France). Here cold and dry conditions prevailed during most of early Aurignacian and were replaced by a warmer, wetter, period for much of later Aurignacian. The general impression is of open landscapes during the cooler stadial and more closed during the warmer interstadial, apparently from cold steppe to warmer parkland environments (Laville *et al.*, 1980) in keeping with the trends described in the previous section. Reindeer frequencies were highest in the early Aurignacian, and red deer, roe deer and wild boar increased in the later Aurignacian, reflecting a greater diversity of local environments (Boyle, 1990) with grassy steppe vegetation plus some arboreal elements, especially conifers in the plateaux and river valleys and thermophilous deciduous trees in the sheltered valleys and south-facing slopes (Wilson, 1975; White, 1985). The situation in the Périgord mirrors the increasingly open habitats with huge biomass and fewer species which were the long-term trends in declining faunal variability in early Upper Palaeolithic (Simek & Snyder, 1988). Blades (1999) concluded that 'a complex mixture of long-term climatic, annual seasonal, and topographic variability influenced the environments of Aurignacian social groups and the fauna they procured.' and that '. . . Aurignacian occupation in Périgord suggests a relationship between environmental structure and diet, at least as reflected in large herbivore fauna.' Crucially in the context of the subject of this section, Blades (1999) found that the proportion of distantly collected materials among the tool portion was greater in the early Aurignacian and the proportion of the tool collection made up of blade tools was also greater. The Aurignacian lithic tool repertoire was considered to be a flexible one that was adjusted in response to the demands of economy and environment (Blades, 1999). A relative increase in burins and thick scrapers in the later Aurignacian may reflect a change in economy. Mobility varied with season and climatic régime. Lithic raw-material variability reflects these adaptations. Adjustments in settlement pattern and mobility were in response to changing distributions of habitats and fauna (Blades, 1999). A similar situation may have occurred in Riparo Mochi, Italy (Kuhn & Stiner, 1998).

Bleed (1986) has provided an interesting alternative way of looking at technology by providing two design alternatives for optimising the availability of any technical system. He distinguishes between reliability and maintainability. Reliable systems are produced so that they are guaranteed to work when needed. Maintenance systems, on the other hand, can be made to function even if damaged or inappropriate to the task being undertaken. Reliable systems characterise populations that exploit resources that are fleeting but predictable by encounter hunting, using special purpose hunting and processing sites. Staged tool production and well-crafted weapons are characteristics of these systems. Maintainable systems, on the other hand, characterise populations that exploit scattered but ubiquitous resources in diverse faunal assemblages by forage hunting. I suggest that reliable systems are characteristic of the Modern Human systems of the open plains that I have described and that maintainable systems are more likely in the case of Neanderthals, or indeed other hominids, exploiting scattered but diverse resources over heterogeneous landscapes.

The intensity of lithic re-use and reduction also reflects features of the landscape and the methods of resource exploitation. Correlates of lithic re-use and reduction intensity include raw material availability, use and rate of wear of different tool types, settlement patterns, mobility and intensity of site occupation, differences in faunal exploitation strategies and prey dispersion (Rolland &

Dibble, 1990). These are 'ecological' correlates that are expected to apply across taxonomic boundaries and should reflect adaptation to particular local circumstances.

Stout (2002), using a present-day ethnographic example of the adze makers of the Langda Plateau in Irian Jaya, makes the point that there is a complex interdependence between social structure and technology. Enhanced capacities for the social facilitation of skill acquisition are defining components of human adaptation. We have here a basis for the burst of technological and symbolic novelty that we observe in the Upper Palaeolithic of Europe that is unrelated to cognitive capacity. The presence of well-developed tool-making skills in archaic humans raises questions about socio-economic context and mental sophistication (Stout, 2002) as I reiterate in this book. The social response to the exploitation of the open plains would have provided a platform for the rapid development and transmission of technological ideas and practices at a faster rate than at any previous time. I would also predict that the rate of innovation and transmission would increase exponentially or in some other non-linear manner. Many of the developments of the Upper Palaeolithic occur some time after the colonisation of the Eurasian Plains and the distinction between late Middle Palaeolithic and early Upper Palaeolithic, as emphasised over four decades ago by Bordes (1961), is not really that great. The fact that such developments did not apply across the board to all Modern Human populations across the world further emphasises the ecologically-mediated social context of the European Upper Palaeolithic.

The establishment, through behavioural (technological) change, of a new carrying capacity or the occupation of ecologically new habitats, which implies a change in resource exploitation, has been proposed by some authors (e.g. Lahr & Foley, 1998) as a possible mechanism that could explain regional expansion of human populations. I am of the opinion that such events would have been rare when crossing regions with dramatically different ecologies (e.g. from the Middle East, across the western Asian mountains, into the North European Plain) and that, in most cases, changing technologies were behavioural responses to new circumstances that would have acted to reduce risk in unpredictable environments with subsequent advantages. Populations with the ability to adjust in this manner would have been at a great advantage. In the case of the Upper Palaeolithic palaeo-indians of North America, a highly portable technology was the optimal solution to a strategy of residential and logistical mobility and frequent home-range change (Kelly & Todd, 1988). The Aurignacian appears to be a similar solution on the Eurasian Plain. Progressive specialisation to large mammal hunting in open plains led to technological developments towards highly efficient and lethal weaponry systems, such as Clovis in North America (Frison, 1989, 1998), microblades in north-east Asia and north-west North

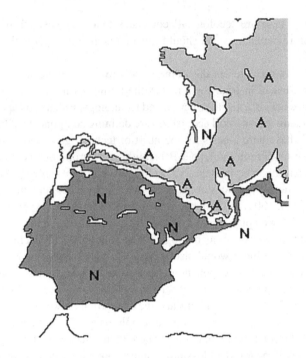

Figure 5.4. Distribution of Aurignacian (A) and Neanderthal (N) populations *c.* 40 kyr in Iberia and south-west France in relation to bioclimate. Dark grey, Mediterranean bioclimates; white, sub-Mediterranean bioclimates; pale grey, temperate bioclimates. Bioclimates after Rivas-Martínez (1996).

America (Goebel *et al.*, 2000), Solutrean in south-west Europe (Straus *et al.*, 2000), Epi-Gravettian in Central Europe (Montet-White, 1994) and possibly, earlier, the Aterian in North Africa (Allsworth-Jones, 1993).

It seems clear that certain technological elements, if not all, had a clear ecological significance and that there was indeed great facultative flexibility in the use of particular tool kits, both among Moderns and Neanderthals. If indeed technology is a reflection of ecology as well as of raw material availability, brain wiring, manual dexterity and social context and, for one moment, we accept that Neanderthals and Moderns were not that behaviourally dissimilar then it has to be accepted that there is no *a priori* reason why technological convergence in response to similar ecological pressures should not happen over and over again. If this is indeed the case we are left with a major problem, that of separating convergent technological evolution from mimicry (or acculturation). The debate will only be resolved if chronology is resolved and there are exponents who will undoubtedly argue the case in either direction at least in the case of south-west

France (d'Errico *et al.*, 1998; Mellars, 1999; Zilhao & d'Errico, 1999, 2000). Perhaps part of the solution to that particular debate is a biogeographical one, one of looking at the problem at the appropriate scale. It is indeed possible, for example, to have Aurignacian 'Moderns' colonising northern Spain before south-west France, especially as the bio-climates of northern Spain are more similar to those of central Europe than are those of south-west France (Figure 5.4). So combining the evidence from these two geographically proximal but bio-climatically dissimilar areas will only serve to compound the problem. I think that the observation that I have made elsewhere (Finlayson *et al.*, 2000a), that the so-called transitional technologies, attributed to Upper Palaeolithic Neanderthals, are distributed on the fringes of the great Eurasian Plain (Figure 7.3) requires an explanation and I think that the most parsimonious explanation may indeed be along the lines of common solutions to common problems.

Symbolic and social behaviour

The Upper Palaeolithic of Europe is associated by many authors with a qualitative jump towards symbolic behaviours, including art, that were previously absent or very rare (Byers, 1994; Ambrose, 2001). The contention is supported by the presence of ornaments made of bone and ivory and parietal art in particular. The implication is that there was a significant difference in the cognitive abilities of the Moderns, who developed these systems, and the archaic hominids, like the Neanderthals, who were incapable of such achievements (Chase & Dibble, 1987; Mithen, 1996; Klein, 1999; Kuhn *et al.*, 2001). Lindly & Clark (1990) claimed that such behaviour was absent from all Middle Palaeolithic humans, not just Moderns, and that the earliest evidence was from the Upper Palaeolithic. However, they sought to decouple symbolism from the biological taxa. I argue here that, as in the case of technology, these observed changes are quantitative rather than qualitative and are the result, not cause, of changing ecological and social circumstances. It follows that explanations that highlight as yet undetected differences in cognition (Klein, 2000) are not required.

Some authors challenge that the abrupt changes in symbolic behaviour claimed for the Middle–Upper Palaeolithic transition reflected cognitive differences and claim to find much earlier evidence of symbolic behaviour. McBrearty & Brooks (2000) summarised the African evidence. d'Errico *et al.* (2001) reported an engraved MSA bone fragment from Blombos Cave in South Africa, and Henshilwood *et al.* (2002) recently reported two abstract representations, older than 65 kyr, engraved in ochre also recovered from MSA layers of this same cave. Riel-Salvatore & Clark (2001) have claimed continuity in mortuary behaviour between the Middle and the Upper Palaeolithic (but see Gargett,

1989). Some authors have suggested even earlier evidence for the origins of symbolic behaviour. Barham (1998) found four pieces of pigment associated with technology at the transition of the Early Stone Age to the MSA in central Zambia. The collection was later enlarged to 302 pieces and this led Barham (2002b) to conclude that pigment use in humans spanned the archaic–modern boundary. d'Errico & Nowell (2000) have also challenged the view of the late emergence of symbolic behaviour with the Lower Palaeolithic Berekhat Ram figurine, a piece of volcanic material from Israel that is claimed to have been purposely modified to produce human features. The Lake Mungo 3 Modern Human, dated at 50 kyr (Thorne *et al.*, 1999; Bowler *et al.*, 2003), was associated with ochre. This last observation seems to support Foley & Lahr's (1997) view that the absence of an Upper Palaeolithic in Australia is not indicative of separation of biological populations prior to the development of modern cognitive capacities. Thus, Modern Human behaviour developed through a process of accretion of new behavioural systems and their biological bases and it follows that there was no sudden appearance of 'modern' behaviour and its antecedents can be mapped through the Pleistocene (Foley, 1992).

I view the increase in symbolism as a reflection of the increasingly sophisticated social systems of the people who entered environments such as those of the Eurasian Plains. A correlation between the abundance of opportunities for social learning and the size of the local cultural repertoire may be an ancient feature, present in the great apes, that may have existed for at least 14 Myr (van Schaik *et al.*, 2003). We have seen how the behaviour of Moderns in the Eurasian Plain would have been correlated with increased opportunities for social learning so it is not unexpected to find a concomitant jump in observed cultural, including symbolic, diversity without recourse for explanations requiring increased cognitive capacity. It is also possible that many of the cultural elaborations observed in the highly unpredictable conditions of the late Pleistocene may be a response to such conditions (Dunnell, 1999; Dunnell & Greenlee, 1999; Hamilton, 1999; Kornbacher, 1999; Madsen *et al.*, 1999).

Humans frequently return to a central place when they forage. Central-place foraging varies along a continuum and involves special travel costs, including carrying time. It also involves deferment of consumption (Kaplan & Hill, 1992). Symbolic communication then becomes important in humans as it is used to increase the information available to foragers in two ways: (a) by increasing the sample size of resource distributions and characteristics; and (b) by providing information on prey or patches that foragers have never exploited (Kaplan & Hill, 1992).

Hunter–gatherers (e.g. !Kung San) maintain kin ties across a region. When resources are scarce people typically respond by visiting kinsmen located in a temporarily well-favoured area (Cashdan, 1992). This may be a kind of reciprocal altruism (Trivers, 1971) that constitutes another kind of buffering in unpredictable environments. Humans may have developed co-operation mechanisms, acting as environmental buffers, by maximising inclusive fitness in kin-based groups (Hamilton, 1964) or through reciprocal altruism (Trivers, 1971) in situations in which the individuals in groups were in regular contact over long time periods. Such co-operation may also evolve among unrelated agents who are sufficiently similar to each other in some arbitrary characteristic (e.g. a marking) but who do not reciprocate (Riolo *et al.*, 2001; Roberts & Sherratt, 2002). Although such co-operation is predicted for organisms with a rudimentary ability to detect environmental signals without need of memory of past encounters, we can see how it could also evolve among large groups of humans with irregular contact. Recent models of large interacting communities of players on spatial grids indicate that voluntary interactions promote co-operation (Szabo & Hauert, 2002). To achieve a stable regime of all-out co-operation other mechanisms, such as punishment, are required (Henrich & Boyd, 2001; Fehr & Gachter, 2002). Image scoring (the way an individual is viewed by the group) also promotes co-operative behaviour in situations where direct reciprocity is unlikely (Wedekind & Milinski, 2000). Whilst the social structure of Neanderthals would have permitted co-operation of a kin or reciprocal type, the larger networks of the Moderns would have permitted a greater range of mechanisms to come into play even among unrelated individuals or those not in regular contact.

I propose that the evolution of strong and complex patterns of social behaviour was a long-term process that had its roots and subsequent development in the increasing adaptation to a plains existence. Such an evolution would have led to increasing group and home-range size that generated the need for within-group division of labour. With this came the need for groups to split for increasingly longer periods. Social bonding and information exchange became increasingly vital. The template for the evolution of a complex social system that would include language, symbolism and art was set in the conditions that were experienced most frequently by Moderns and less so, because of habitat differences, by archaics such as the Neanderthals.

The fact that Neanderthals lived in the heterogeneous landscapes of the Mediterranean bioclimates meant that they were able to obtain sufficient resources within small home ranges (Mellars, 1996; Gamble, 1999). This system, which was linked to a highly omnivorous diet, did not lend itself to the social innovations that were a feature of the peoples of the plains (Gamble, 1999). Anderson & Franks (2001) distinguish between groups and teams in animal societies. Groups are associated with small societies in which group tasks require the simultaneous performance and co-operation of two or more individuals for successful completion. Teams, on the other hand, are more likely to be found in larger, more complex, societies with a division of labour. Specialisation, learning and overall enhanced performance efficiency enhance the ergonomic efficiency of the team (Anderson & Franks, 2001). In this respect I see Modern societies behaving more like teams, based on individual recognition, acting as cliques with an element of trust among individuals and exposed to potential cheats. Neanderthal societies may have behaved more like groups. Neanderthals probably lived most of their lives in small family groups, in which the individuals would have been in close contact throughout their lives, within an intimately well-known home range, and with little within-group division of labour. This does not imply any Modern superiority. It simply indicates that these were adaptations to improve efficiency in wide-ranging people, adaptations that may not have been necessary in the environments exploited by the Neanderthals.

Language

Lieberman (1989) suggested that Neanderthals were deficient with respect to their linguistic and cognitive capacity, their speech communications being nasal and subject to perceptual errors. Stringer & Gamble (1993), discussing the probability that Neanderthals had language, concluded that the Neanderthals did have at least a rudimentary form of language but that it was probably simple in construction and restricted in its range of expression. These authors recognised, however, that the ability to create and understand language was a function of the brain rather than of the vocal tract, which presumably they considered to be related to speech. The lack of understanding of the Neanderthal brain was recognised as a problem. Aiello & Dunbar (1993) considered that by 250 kyr the hominid brain had reached a size that could sustain group sizes of 150 people and that such group sizes required a gossip (instead of a full symbolic) language. Buckley & Steele (2002), however, did not find a link between the management of affiliative ties in large co-residential social groups and language. Instead, they considered that life history strategy was the prime mover, within the context of co-operative foraging and provisioning, for spoken language abilities. Speech adaptations, according to this view, co-evolved with life history strategy.

Language, as a social adaptation, allows individuals to transfer and receive information about people, objects and places not present, and to manage relationships with other individuals (Buckley & Steele, 2002). Narrative devices would have served to communicate foraging knowledge (Sugiyama, 2001). Gamble (1999) considered that Neanderthals had an advanced form of communication that was, however, only directed towards the negotiation of their intimate networks. In a comparative study of Middle and Upper Palaeolithic technology, Chazan (1995) could find no evidence to support the hypothesis that the transition was related to the development of language. Ohnuma *et al.* (1997), on the other hand, inferred experimentally that spoken language was not indispensable for Levallois flake production in the Middle Palaeolithic. In my view it is inconceivable to think of Neanderthals without language. Recent work with bonobos *Pan paniscus* reveals the ability of these apes to work with a, limited, language (Savage-Rumbaugh & Lewin, 1994). Tonal melodies form musical gestalts in rhesus monkeys, as they do in humans, indicating that there is similar transduction, storage, processing, and relational memory of musical passages in monkeys and humans (Wright *et al.*, 2000). Large-brained hominids would have had far greater capacities, especially the Neanderthals. This does not imply that Neanderthals and Moderns had identical forms of language and speech. Buckley & Steele's (2002) explanation for the evolution of language via intensely negotiated co-operation within small stable groups based on family and kinship ties would fit the Neanderthal model and thus advocates a pre-Modern evolution. The different lifestyles of Neanderthals and Moderns and the manner in which Moderns expanded their scale of operation, with temporary absences of individuals from the group becoming a regular feature, would have placed a strong pressure on the refinement and further development of any language and speech capability that existed in archaic forms. The dating of the fixation of the mutation in the FOXP2 gene, that is involved in the developmental processes that culminate in speech and language (Lai *et al.*, 2001), to within the last 200 kyr (Enard *et al.*, 2002) is in keeping with this scenario. These refinements could evolve by a Baldwin effect (Simpson, 1953), language emerging through self-organisation and continued through cultural transmission. Individual characteristics that were advantageous to language production and comprehension would then be fixed by natural selection (Buckley & Steele, 2002).

Neanderthal–Modern ecological and behavioural differences

What conclusions can we draw with regard to differences in behavioural ecology between Neanderthals and Moderns? The main conclusion is that the two forms were highly intelligent and behaviourally plastic and capable of responding dynamically to ecological circumstances. The dehumanised portrayal of the Neanderthals is not consistent with the evidence of a sophisticated hominid having the root of many of these behavioural patterns in pre-Neanderthal populations (Hayden, 1993). It is therefore not surprising to find that there is so much overlap in the food consumed and the way in which it was procured.

Neanderthals were well adapted to the nature of the terrain that they exploited. Geographically, it was the MLB from Portugal in the west to the mountains of Central Asia in the east (Chapter 3). This was the geographical area of continuous occupation (Gamble, 1999; Finlayson *et al.*, 2000a) of the Neanderthals and their predecessors (Figure 5.3). So for a period from the establishment of the pre-Neanderthals, perhaps around 500 kyr (Chapter 4), until their extinction around 30 kyr they adapted to the heterogeneous landscapes of this MLB, expanding northwards at the western end of the range when milder conditions removed the open steppes from north-west Europe (Chapter 6). I argue here that the robust morphology of the Neanderthals, typical of Middle Pleistocene hominids, suited them well but also prevented niche expansion to include open plains environments that required large annual home ranges.

In the mosaic landscapes of the Levant the Moderns occupied the range of available habitats but their mobility regime appears to have been very different from that of the Neanderthals, exploiting resources seasonally, implying a greater annual home range. In the Levant we observe a natural laboratory in action. We have two morphological types of human, behaviourally similar and plastic but exhibiting differences in the way they utilised the landscape. Since both were regionally present for a period of up to 60 kyr (from 100 to 40 kyr ago) we can infer that climatic and environmental fluctuations, sometimes favouring one strategy, at other times the other, prevented one form succeeding over the other. We would expect that similar patterns would be observed wherever the two forms were present in terrain that included heterogeneity and plains, as in the Levant. Another such area would be the northern edge of the central Asian mountains with the Russian Plain (Soffer, 1985, 1994). The difference here is the much greater extent of the plains and the huge herbivore biomass resource available. Such a situation awaited a hominid that could break away from the shackles of the territorial exploitation patterns that had characterised Eurasian Middle Palaeolithic hominids. It is unlikely that the Neanderthals would be the ones to suddenly exploit this resource, which would have been at their doorstep for millennia. Gene flow between the populations in the heterogeneous belt and hypothetical breakaway populations on the plains would have limited morphological change, the necessary base for the successful exploitation of the plains.

A similar situation occurred to the Neanderthals in the west, in areas such as south-western France, Italy and around the Carpathians. These were edge areas between heterogeneous landscapes and the plains. During the cold and arid events of OIS 3 (Chapter 6) the steppic environments expanded westwards. Neanderthals in these edge areas had to adapt to the changing circumstances or face local extinction. The evidence seems to point to behavioural adaptation in the form of transition industries (d'Errico *et al.*, 1998; Mellars, 1999) that appear to reflect changing hunting tactics and even an increase in home range (Feblot-Augustins, 1993). Morphology could not change rapidly enough and the Neanderthals disappeared after a few thousand years.

Returning to central Asia, we have no evidence of major climatic or physical barriers that would have separated plains Middle Palaeolithic hominids from those in the adjacent heterogeneous belt. Instead, we have a morphologically gracile candidate in the area as from 100–60 kyr, with an ability to exploit plains environments. Although we cannot fully discard a local plains adaptation by Neanderthals, as might be expected under stressful situations in the presence of resource heterogeneity (Parsons, 1993), my view is that it was the early Moderns that penetrated into central Asia, operating at larger spatial scales and evolving increasingly sophisticated behavioural patterns in ecotonal areas between the central Asian belt and the Russian Plain. The territorial exploitation system and its correlates, that included living in large groups, intense social binding mechanisms and mechanisms of food storage would have provided the launch pad for the conquest of the Russian Plain. Morphology was no longer a constraint. Behavioural adaptations, not dependent on slow genetic change, catapulted the colonisation. This process is in keeping with the observation that innovation of behavioural traits is especially likely during periods of stress (Parsons, 1993). That events during the last glacial cycle introduced stress (Chapter 6) is beyond doubt. Behavioural plasticity, allowing for changes in social behaviour, may be widespread in animals (e.g. Davies, 1976) and may be the optimal response to stresses that occur at higher frequencies than generation time. We detect this as the rapid spread of the Aurignacian people across the Eurasian Plains that I view as a classic form of ecological release (Cox & Ricklefs, 1977). I will examine this expansion in detail in Chapter 7 after I have described the climatic and environmental changes that affected the world at this critical time.

Kaplan *et al.* (2000) have proposed four characteristics of humans: exceptionally long life span; extended period of juvenile dependence; support of reproduction by older post-reproductive individuals; and male support of reproduction through the provisioning of females and their offspring. These attributes, along with extreme intelligence, are co-evolved responses to a dietary shift toward high-quality, nutrient-dense, difficult-to-acquire resources. If this is indeed the signature throughout the course of human evolution, we would expect its maximal expression in those hominids that ventured onto open plains environments and specialised on large mammalian herbivore meat.

Group size is a function of the size of the neocortex in primates (Dunbar, 1992). The relationship may be the result of limits imposed by the number of neocortical neurons on the information-processing capacity of organisms that would set a cap on the number of relationships that an individual can monitor simultaneously. Since group size will be determined by the ecological characteristics of the environment occupied, new environments that require larger groups will only be successfully colonised after the evolution of larger neocortices (Dunbar, 1992). The recent discovery of microcephalin, a gene that is expressed in the developing cerebral cortex of the foetal brain, may provide insights into the proximate mechanisms of neocortex development and evolution (Jackson *et al.*, 2002). The enlarged neocortex of Moderns when compared to archaic forms (Stringer & Gamble, 1993) may be an additional feature that promoted the rapid expansion into plains environments. In this scenario, the observed behavioural characteristics of Upper Palaeolithic Moderns on the Eurasian Plain would reflect a quantitative, but not a qualitative, catapulting of this co-evolutionary process.

The time frame for the emergence of uniquely human cognitive skills is between 2 Myr and 0.3 Myr, probably closer to the youngest age (Tomasello, 1999). According to this author the emergence may have been marked by a significant genetic event but one that did not specify the detailed outcomes we see in humans today. Instead, it opened the way for new social and cultural processes that subsequently, with no further genetic events, created the distinctive features of human cognition. In my view, and in keeping with Tomasello's (1999) time frame, the 'significant' event probably pre-dated the Modern–Neanderthal split. This would have permitted a multiplicity of subsequent outcomes. When I compare the Neanderthals and the Moderns I can only come to one conclusion – they expressed two alternative forms of being human. It is only the difference in morphology that, in the end, marked the difference by restricting or releasing behavioural patterns that ultimately permitted the survival of one and the extinction of the other in a randomly fluctuating world.

Synthesis

In the previous chapters we have seen how African humans evolved and colonised non-tropical areas of the Earth, facing increasing uncertainty in the process. There was a split between the African and Eurasian lineage around 500 kyr. By 120 kyr we recognise a Modern Human lineage in Africa and the Neanderthals in Eurasia. Morphological differences characterised the two lineages. In this chapter we have compared the behavioural ecology of the Neanderthals with that of the colonising Moderns and concluded that the differences between them are largely facultative, reflecting ecological circumstances. The rapidly fluctuating environments of the late Pleistocene conferred advantages to either form at different times permitting a regional 'coexistence' for millennia. We saw in Chapter 4 that behavioural flexibility was the solution to increasing instability. Neanderthals resolved it by living in heterogeneous, species diverse, environments. Moderns did so by incorporating mammal herbivore-rich plains and increasing their scale of operation. In this chapter I have refined the behavioural characteristics of Neanderthals and Moderns that may be summarised as follows.

Neanderthals

Neanderthals continued the exploitation of mammalian herbivore meat as a mechanism of enhancing flexibility to cope with periodic food uncertainties. The heterogeneous environments of the MLB, proximity of coastlines, abundance of wetlands and rocky habitats and a relatively southerly latitude permitted a degree of omnivory as an additional solution for coping with uncertainty. Neanderthals were opportunistic and they hunted and scavenged. Close quarter hunting was practised in the intermediate habitats and landscapes that they occupied. Their robust morphology was compatible with this method of hunting. Neanderthal expansion onto the Eurasian Plains was, however, limited because of this morphology. Technology reflected the ecological circumstances described above. Attempts to develop technology suited for open habitats and landscapes are observed in edge regions as open environments encroached. Neanderthal tools are largely of the maintenance type. Neanderthals operated as groups in the landscape. Language and symbolic behaviour would have been of a kind that would have sufficed the needs of individuals that were in regular contact with each other during the life cycle.

Modern humans

Moderns also continued the exploitation of mammalian herbivore meat as a mechanism of enhancing flexibility to cope with periodic food uncertainties. In regions with heterogeneous environments as well as coastal areas, wetlands and rocky habitats omnivory was an additional solution for coping with uncertainty. Moderns, like Neanderthals, were opportunistic and they hunted and scavenged. The morphological gracilisation of the Moderns permitted expansion onto the Eurasian Plains. Exploitation of open plains environments was favoured by cold and aridity. Food would have been patchily distributed on the plains. Correlates of reducing risk and successfully colonising open plains were:

(1)   almost total dependence on mammalian herbivore meat;
(2)   fat deposition;

 (3)   storage;
 (4)   food sharing;
 (5)   trade;
 (6)   large home, annual and inter-annual ranges;
 (7)   large group size;
 (8)   within-group division of labour and operation as teams;
 (9)   reliable-type technology;
(10)   long-range, projectile, technology;
(11)   tool re-use;
(12)   complex social structure including kin- and non-kin altruism;
(13)   rapid development and transmission of technological ideas;
(14)   behaviours that transmitted information not directly experienced by all
       group members – symbolism and complex spoken language.

These behaviours were present in archaic hominids but were far more promi-
nent in plains dwelling Moderns. The behaviour of Moderns therefore evolved
gradually by a process of accretion and was stimulated by a plains existence.
Inventive behaviour was maximised in these situations of uncertainty. These
stress-induced behavioural innovations resolved stresses that occurred at frequencies higher than generation time that could not be matched by genetic
adaptation. The Moderns in the Eurasian Plains thus behaved cohesively as a
super-organism. It was the key to dealing with risk.

6 The conditions in Africa and Eurasia during the last glacial cycle

The global pattern

In this chapter I will focus my attention particularly on temperate and boreal Europe, the Mediterranean and Africa, which are the key areas that will be examined in Chapter 7 with regard to the Neanderthal extinction and the colonisation by Moderns. In Chapter 3 I described the pattern of increasing global climatic deterioration and instability during the Pleistocene. This progressive deterioration led, for example, to the contraction and extinction of tropical and sub-tropical woodland in southern Europe and to the rise of xeric species that culminated with a maximum expansion during the Last Glacial Maximum (LGM) (Carrion *et al.*, 2000). Smaller scale patterns, as will be discussed in this chapter, have to be viewed within these larger-scale climate trends at the scale of millions of years (Webb & Bartlein, 1992). The progressive glaciation of the northern hemisphere commenced towards the end of the Pliocene although cooling started as early as the Eocene. The Quaternary is characterised by the alternation of cold glacial and warmer interglacial periods. There were at least nine glacial–interglacial cycles between 2 Myr and 700 kyr (Shackleton & Opdyke, 1973, 1976; Shackleton *et al.*, 1984) with at least 10 after that (Imbrie *et al.*, 1984). Interglacials, which were often brief, started and finished abruptly (Flohn, 1984; Broecker, 1984), and characterised only 10% of the Pleistocene (Lambeck *et al.*, 2002a & b). The amplitude of the climatic oscillations was lower prior to 735 kyr (*c.* 41 kyr) than after (*c.* 100 kyr) (Ruddiman *et al.*, 1986). During glacial events, sea levels dropped to between 90 and 130 m below current levels (Shackleton & Opdyke, 1976; Rohling *et al.*, 1998; Lambeck *et al.*, 2002a). Superimposed on these cycles were shorter episodes of ice expansion during interglacials (stadials), when conditions were cold and arid, and of thermal improvement and increased humidity during glacials (interstadials) (Voelker, 2002); at lower latitudes cycles of rainfall (pluvials) and aridity (interpluvials) were typical. Transitions were often rapid, especially from glacial to interglacial conditions (Webb & Bartlein, 1992). Global warming at the start of the last interglacial occurred at the rate of 5.2 °C/1 kyr (Ruddiman & McIntyre, 1977). Woillard (1979) calculated that northern French temperate forests were replaced by pine, spruce and birch taiga at the end of the last interglacial in the

135

space of between 75 and 225 years. Dansgaard *et al.* (1989) have shown that temperatures rose by 7 °C in 50 years at the end of the last glaciation. Stuiver & Grootes (2000) record 13 cold–warm transitions in the period 60–10 kyr that took 50 years to complete, the initial response being very fast with only a few years being required to reach the mid-point.

In this chapter I focus on the last glacial cycle (125–0 kyr) that has been the coolest and most variable period and that which saw the arrival of Moderns into Eurasia and the extinction of the Neanderthals. It was characterised by the climate variations that were typical of the late Pleistocene, being greatest in amplitude and severity during glacial maxima (Oppo *et al.*, 1998) when atmospheric circulation was most turbulent (Ditlevsen *et al.*, 1996). The last glacial cycle was characterised by the melt of Oxygen Isotope Stage (OIS) 6 ice around 130 kyr, followed by a brief interglacial (OIS 5e) and a subsequent gradual climatic deterioration at OIS 4 (the ice sheet being smaller than in the LGM). This was followed by a long drawn ice withdrawal ahead of the LGM in OIS 2 (van Andel & Tzedakis, 1996) when winter sea ice extended south to the French coast (COHMAP, 1988). The intermediate OIS 3 (60–25 kyr) is characterised by many climate changes at millennial scales, with cold periods that approached conditions at the LGM and warm intervals with temperatures just below Holocene values (Barron & Pollard, 2002). Large areas of Fennoscandia were, however, ice-free as the Scandinavian Ice Sheet was much reduced in comparison with OIS 2 (Ukkonen *et al.*, 1999; Arnold *et al.*, 2002).

Global climate has also fluctuated repeatedly and frequently even within glacials and interglacials (van Andel & Tzedakis, 1996; Nimmergut *et al.*, 1999; Watts *et al.*, 2000) although the degree of climatic variability of the last inter-glacial is disputed (Kukla, 1997, 2000; Kukla *et al.*, 1997; Cheddadi *et al.*, 1998; Frogley *et al.*, 1999; Rose *et al.*, 1999; Boettger *et al.*, 2000). OIS 4 appears to have been more climatically diverse than had been recognised and included interstadial conditions (Watts *et al.*, 2000). The GRIP ice core data (Dansgaard *et al.*, 1993; GRIP, 1993) reveals 20 warm events in the time period 105–20 kyr. Such large and abrupt, high frequency, climate changes seem to have occurred globally at annual, decadal, centennial and millennial intervals (Bond *et al.*, 1997; Allen *et al.*, 1999; Alley *et al.*, 1999; Alley, 2000; Stuiver & Grootes, 2000; Elliot *et al.*, 2002; Helmke *et al.*, 2002; Voelker, 2002), particularly during OIS 3 (van Andel & Tzedakis, 1996). The Dansgaard–Oeschger (DO) temper-ature oscillations (Dansgaard *et al.*, 1993) are the dominant millennial-scale climate-change signal and are associated with the alternation between warm times and glacial maxima and stadials in the North Atlantic (Alley *et al.*, 1999). Heinrich Events (HE) are less frequent and are the result of massive ice dis-charges into the North Atlantic related to the surging of the Laurentide Ice Sheet through the Hudson Strait (Heinrich, 1988; Bond & Lotti, 1995; Alley *et al.*, 1999). There were, additionally, brief warm events such as Oerel (58–54 kyr), Glinde (51–48 kyr), Hengelo (39–36 kyr) and Denekamp (32–28 kyr) intervals. These oscillations caused repeated advances of forest and open vegetation in Europe and the Mediterranean and of Mediterranean forest and semi-desert in north-west Africa (van Andel & Tzedakis, 1996).

Environmental change, indicated by vegetation dynamics, not only reflects responses to global events translated into local climatic changes but is also a function of the available pool of plant taxa, thus creating spatial and temporal heterogeneities of response (Tzedakis, 1994; Vandenberghe *et al.*, 1998). For example, Gibraltar is a coastal site that is separated from the interior by high coastal ranges. Here cooling is reflected by the presence of montane black pines *Pinus nigra* (Finlayson & Giles Pacheco, 2000) and not by the presence of the open vegetation that characterises inland sites (Carrión, *et al.*, 1998). Chance changes in the location of plant glacial refuges are critical in the subsequent recolonisation of adjacent areas during interglacials (Reille & Beaulieu, 1995). Time lags are also crucial in the interpretation of data. Care has to be exercised when applying terrestrial data to the marine record. Such assumptions may be valid in the case of transitions from glacial to interglacial conditions that are rapid and large scale. Interglacial to glacial transitions are, however, more diffuse because ice-sheet growth is slower than ice-sheet decay. In such cases vegetation may respond differently and at different rates leading to time lags between vegetation and climatic variables (Tzedakis, 1994; Coope, 2002). Time lags are not only important in the case of vegetation change. Faunal patterns, including those of humans, are also subject to time lags and this can cause confusion and misinterpretation of cause and effect, particularly at small temporal scales. Direct correlations of faunal patterns to climate signals from the marine isotope record therefore need to be addressed with caution.

Sea-level changes in the Quaternary have been the result of the growth and decay of the ice sheets. Sea-level lowstands are associated with glaciations and highstands with interglacials. The global picture is of a rapid rise in sea-levels, due to an exceptionally fast OIS 6 deglaciation, of the order of 20 m per 1 kyr to a highstand during OIS 5e with sea-levels between 2 and 12 m above the present (van Andel & Tzedakis, 1996). This was followed by a variable but progressive drop in sea-levels to the LGM lowstand of between -118 and -135 m (Clark & Mix, 2002), levels oscillating around -80 m during OIS 3 (Barron & Pollard, 2002) and falling rapidly from around 30 kyr with a sharp subsequent rise to the Holocene highstand (van Andel & Tzedakis, 1996). Prior to the LGM there were significant and rapid sea-level fluctuations of the order of 10 to 15 m every 6 kyr, coinciding with Heinrich ice-rafting events (Lambeck & Chappell, 2001). There were then two major periods of rapid and sustained sea-level rise

(16–12.5 kyr and 11.5–8 kyr) at a rate of 15 m per 1 kyr (Lambeck & Chappell, 2001).

Superimposed on the global signals are regional and local changes that have been caused by uplift and subsidence of the coastal zone or by regional and local climate effects. Climatic, meteorological and tidal-driven changes are particularly important at decadal, annual and even smaller scales and may lead to significantly different patterns even between proximate locations (Lambeck & Chappell, 2001).

Temperate and boreal Europe: the Eurasian Plain

The colonisation by trees of the Eurasian Plain during interglacials reflected differences in source areas of individual species, different rates of migration and also in the nature of the climatic signal so that the broad-leaved forest succession varied regionally and between different interglacials (Zagwijn, 1992). In the last interglacial (OIS 5e) much of Europe was covered by temperate mixed oak forest (Vandenberghe *et al.*, 1998; Turner, 2002; Kukla *et al.*, 2002) contrasting with the great reduction of forest and replacement by treeless vegetation types during glacials (Reille *et al.*, 2000; Kukla *et al.*, 2002). This was an oceanic interglacial, a rare form that only occupied 12% of the time span of the last 500 kyr and was characterised by a constant succession in the expansion of elm *Ulmus*, oak *Quercus*, hazel *Corylus*, yew *Taxus* and hornbeam *Carpinus* followed by fir *Abies*, spruce *Picea* and pine *Pinus*. Tree response was less intense during continental warm episodes (Zagwijn, 1992).

In contrast, at the height of the LGM (OIS 2) in Europe, around 20 kyr, glaciers covered around two-thirds of the British Isles, Scandinavia, the Baltic and the Alps. Much of the remainder of Europe north of the Mediterranean had a cover of permafrost (CLIMAP, 1976; Maarleveld, 1976). Away from the ice sheets the vegetation cover was sparse with large areas of bare ground and an unusual mixed flora (Birks, 1986; Zagwijn, 1992). Much of Europe was a polar desert with varying local conditions of geology, soil, relief and microclimate contributing towards variations in the local flora. The ice sheets were even more extensive and persisted longer during the cold phase (OIS 6) preceding the last interglacial (OIS 5e) (van Andel & Tzedakis, 1996).

The situation during OIS 3 suggests that, even in oceanic temperate areas such as Britain, mean July temperatures between 40 and 25 kyr were close to 10 °C (temperatures that limit tree growth) and winter temperatures were low, around −20 °C. These estimates are based on the presence of cold, tundra, insects in British deposits, including a number of species that are currently restricted to east Asia and indicate that a treeless landscape with a tundra flora and fauna predominated (Coope, 2002). Evidence from other sites across the North European Plain indicates that a uniform climate extended across much of this region at the time. The vegetation of the Eurasian Plain during OIS 3, in the west at least, seems to have oscillated between conifer woodland, with shrub tundra in the north, during warm parts of the cycle and a tundra and cold steppe mosaic, with polar desert in the north, during the coldest moments (van Andel & Tzedakis, 1996). Coope (2002) proposed three modes for the OIS 3 climate of north-western Europe: (1) a brief period between 43 and 42 kyr of temperate and oceanic conditions, with similar temperatures to today but an absence of trees in the landscape; (2) a series of interstadials between 40 and 25 kyr with cold continental conditions as described above; and (3) periods of intense cold between the interstadials during which biological systems were reduced to a minimum or inhibited altogether.

Further east, on the Russian Plain, the dense temperate forest of the last interglacial maximum was replaced by open, harsh, loess-steppe with the onset of the LGM (Rousseau *et al.*, 2001). The latter part of OIS 3 is characterised by decreasing temperature, increasing aridity and continentality (Soffer, 1985). Extremely cold temperatures, aridity and permafrost characterised large parts. The vegetation was a unique tundra-grassland with sparse arboreal vegetation (pine, beech, oak) in the form of gallery forests along river valleys (Soffer, 1985). Conditions improved in relative terms between 33 and 24 kyr during the Briansk Interstadial before the onset of full glacial conditions (Soffer, 1985; Markova *et al.*, 2002). Tundra, forest–tundra and tundra–steppe were widely distributed across the Russian Plain during the Briansk Interstadial with the southern limit of the range of a number of Arctic plants being 1200 km further south than today (and a further 600 km during OIS 2). Steppe and forest–steppe predominated in the south, for example around the Black Sea. In the south, the added topographical heterogeneity (e.g. in Crimea) increased the diversity of vegetation types (Markova *et al.*, 2002). Climatic improvement is also detected in northern Siberia from 48–25 kyr, with open larch forest with *Alnus fruticosa* and *Betula nana* in the Taymyr Peninsula (Andreev *et al.*, 2002).

In contrast to the northward migration of trees during warm episodes the effect of glacials was the extinction of trees except within glacial refugia (Willis, 1996; Tzedakis *et al.*, 2002). By around 13 kyr thermal conditions had improved in north-western Europe and the tundra and steppe were gradually replaced by boreal woodland and then by spruce forest and by birch–conifer woodland (Huntley & Birks, 1983). The cold Younger Dryas Stadial (11–10 kyr) represented a further deterioration of conditions with tundra once again stretching from southern Sweden to much of France and the British Isles (Huntley & Birks, 1983). The rapid amelioration leading to the present interglacial (the Holocene) followed after 10 kyr BP. These changes are reflected in sites with long pollen sequences. At Grand Pile in France the long sequence spanning the past 140 kyr records significant changes in arboreal and herbaceous pollen that can be correlated with marine isotopic climate signals (Woillard & Mook, 1982).

Sites on the southern fringe of the Eurasian Plain and bordering the Mediterranean lands may reflect the climatic and environmental changes of the glacial–interglacial cycles more accurately than those within the Mediterranean or of the Eurasian Plain. In areas with sharp ecotones vegetation responses are sensitive and rapid because there is very little migration lag as all the response species are present within the geographical area, forming a vegetation–climate mosaic at any given time (Blasi *et al.*, 1999; Peteet, 2000; Roucoux *et al.*, 2001). At Lago Grande di Monticchio, an Italian lacustrine sequence spanning the past 102 kyr, the mean interval for absolute changes of >20% in total pollen of woody taxa was 142 yr with decreases being more rapid than increases (Allen *et al.*, 1999). French Massif Central sites such as Lac du Bouchet and the Praclaux Crater, because of their altitude and location relative to refuges and also the altitudinal vegetation zonation, detect low amplitude climatic fluctuations better than Mediterranean sites in which refugia persisted throughout glacial episodes (Reille & Beaulieu, 1995; Reille *et al.* 1998). The contrast is a reflection of the reality of the division between the Eurasian Plain and the mid-latitude belt that I have stressed throughout this book. The Eurasian Plain would have experienced wider environmental swings in response to climate change than more southerly areas, not just for reasons of latitude but also on account of refugial persistence of species. Edge areas, not just on the fringes of the European peninsulas and the Eurasian Plain but also on similar areas on the edge of the Russian Plain (Markova *et al.*, 2002) would therefore have exhibited huge temporal and spatial ecological diversity.

The Mediterranean

Conditions in the Mediterranean would not have changed in the relatively simple manner described for temperate and boreal Europe. The vegetation of the Mediterranean would have been controlled, as it is today, by the geography of the landscape and the local climatic peculiarities of each area (Suc *et al.*, 1994). The longitudinal width of the Mediterranean and the west–east orientation of the major mountain masses are barriers for plant movement. High ground extends far south in peninsulas, particularly in Iberia, the Balkans and Italy, and this permitted the intrusion of some elements of temperate vegetation well into the Mediterranean bioclimatic zones (Rivas-Martínez, 1981, 1987; Zagwijn, 1992). The double seasonal climatic rhythm is today highly heterogeneous depending on the variable influence of Atlantic air, desert conditions and local relief.

The relative influence of wet and arid cycles will have varied during glacial–interglacial cycles (Narcisi, 2001). In southern Europe, in particular, moisture is a critical ecoclimatic variable with temperature playing a supporting role (Tzedakis, 1994) and precipitation was a limiting factor to many plants during glacials (Willis, 1996). The southward displacement of a weakened Gulf Stream (Lynch-Stieglitz *et al.*, 1999) to the shores of Portugal (van Andel & Tzedakis, 1996), may nevertheless have at times ameliorated glacial conditions in southwestern Iberia. At other times Heinrich events would have significantly cooled these areas (Broecker & Hemming, 2001). Many areas of the western Mediterranean would have experienced harsh conditions during glacials and stadials (Rose *et al.*, 1999) reflecting the spatial mosaic characteristic of the region. The low latitudinal situation would have additionally permitted significant diurnal warming, especially in the summer, even during cold phases. The gradual shift from peak interglacial to early glacial from high to middle latitudes (Kukla *et al.*, 2002; Shackleton *et al.*, 2002; Tzedakis *et al.*, 2002) is a further indication of the relatively benign conditions of the Late Pleistocene Mediterranean in comparison with the Eurasian Plain (Prokopenko *et al.*, 2002).

Patches of Mediterranean vegetation therefore persisted even during the coldest and most arid phases and these patches would have varied in distribution and size in relation to local variations of temperature and humidity (Florschutz *et al.*, 1971; Pons, 1984; Reille, 1984). These southern refuges maintained a significant plant diversity that permitted periodic expansions during interstadials (Carrión *et al.*, 2000). The episodic contraction of the geographical range of Mediterranean woodland taxa to southern intra-montane and coastal refugia in response to climatic deterioration is a feature of the Pleistocene of Iberia (Carrión *et al.*, 2000). In Gibraltar the presence of olive *Olea europaea*, a species considered to be an indicator of maximum interglacial conditions (Tzedakis, 1994; van Andel & Tzedakis, 1996), virtually throughout the sequence spanning the last interglacial to the present (Finlayson & Giles Pacheco, 2000) indicates the refugial nature of southern coastal sites. Inland, climatic fluctuations varied the extent of tree cover, dominated by *Pinus*. The coldest and most arid periods favoured steppe vegetation but Mediterranean taxa persisted. Woodland replaced open vegetation with climatic warming. The last interglacial, with mean annual temperatures of around 2 °C higher than the present, saw the development of extensive woodland and the maximal expansion of olive and evergreen oak across the Mediterranean (Tzedakis, 1994; Rose *et al.*, 1999). Forest development during interglacials, however, appears to cover only a fraction of the entire period (Tzedakis, 1994). These patterns are similar in other parts of Mediterranean Iberia with oscillations in vegetation cover from woodland to open vegetation and even a breakdown of vegetation cover (Rose *et al.*, 1999), in the relative abundance of thermophyllous species, and in the

Figure 6.1. Present distribution of thermo-Mediterranean bioclimate (white) in relation to other Mediterranean (grey) and Euro-Siberian (black) bioclimates. After Rivas-Martínez (1981, 1987).

alternating development of broad-leaved and coniferous woodland (Carrión *et al.*, 2000).

The development of Mediterranean vegetation and mixed forest during OIS 3 has been observed in a number of Iberian Mediterranean localities (Burjachs & Julia, 1994; Carrión, 1992; Carrión *et al.* 1995; Carrión & Munuera, 1997). In Italy the forest expands during warm phases but never to the extent reached during an interglacial, creating a mosaic landscape of forest and grassland (Watts *et al.*, 2000). The extent and location of the Iberian refugia were probably much greater than currently described in European maps based on limited Iberian pollen sources (van Andel & Tzedakis, 1996, 1998). The evidence instead suggests that there would have existed a large refugium within the areas currently occupied by the thermo-Mediterranean bioclimatic zones (Figure 6.1). The apparently contrasting evidence of a succession of cold and temperate environments in the Iberian Peninsula between 50 and 30 kyr (Sanchez Goñi *et al.*, 2000a) is easily reconciled. The location of the marine core, off the coast of Lisbon, strongly indicates that it is sampling material preferentially derived from the continental central mesetas of the Iberian Peninsula, that would characteristically have exhibited an alternation of deciduous and evergreen oak woodland with steppic vegetation and periods with the virtual elimination of Mediterranean vegetation, and the Atlantic Portuguese coast that even today has a Euro-Siberian vegetation component (Rivas-Martínez 1981, 1987). The environments characteristic of the Mediterranean glacial refugia would have been under-represented or not represented at all in such a core (Figure 6.1). It is therefore not surprising either that such a sequence should resemble other continental Mediterranean sites such as Lago Grande di Monticchio (Allen *et al.*, 1999; Sanchez Goñi *et al.*, 2000a). To the north, north-western Iberian patterns of vegetation change between 65 and 9 kyr, alternating between herbaceous vegetation with small tree refugia during stadials and discontinuous woodland during interstadials (Roucoux *et al.*, 2001). These differences over relatively short distances emphasise the heterogeneous nature of the Iberian Peninsula (Finlayson *et al.*, 2000a).

There is evidence that in southern Iberia the marine fauna was more sensitive to climate change than the terrestrial fauna. Levels associated with the end of OIS 3 in the Gibraltar sites have produced a record of North Atlantic and Arctic marine mammal (Atlantic grey seal *Haliochoerus gryphus*) and bird species (long-tailed duck *Clangula hyemalis*, little auk *Plautus alle*, great auk *Alca impennis*) that are nowadays rare or absent from these latitudes (Finlayson & Giles Pacheco, 2000). Such incursions may reflect southern extensions of polar water and the presence of icebergs off Portuguese waters on at least six, Heinrich event-related, occasions between 65 and 9 kyr (Roucoux *et al.*, 2001).

In Greece, Tzedakis (1994) recognises two orders of change in vegetation, a pattern that is probably typical throughout the Mediterranean. There is one at the level of open, herbaceous, to forest vegetation that reflects glacial–interglacial cycles. The other, of lower order, reflects changes due to forest succession and in the character of open vegetation. Importantly, this author recognises that between glacials and interglacials there are long periods, taking up between 70 and 80% of the cycle, that are intermediate in nature. In his study of the Ioannina 249 core from Greece, Tzedakis (1994) found that these intermediate periods were characterised by steppe–forest, forest–steppe and steppe vegetation. The extremes were characterised by desert–steppe or forest. In Italy open and arid environments were also characteristic of glacial phases, with less open or closed but humid environments during interglacials (Montuire & Marcolini, 2002). A study of the micromammal fauna of Italy revealed similar patterns of climate and environmental change in the north and in the centre–south. Nevertheless, conditions were always more temperate in the centre–south indicating that there may have been areas that acted as refuges for micromammals (Montuire & Marcolini, 2002).

In Greece, as probably over much of the Mediterranean, cold stages are not uniform within. Instead such periods are characterised by a shifting balance of open vegetation types, always with a relative abundance of relict tree populations within the landscape (Tzedakis, 1993; 1994). Interglacials also appear to have been variable and composed of smaller-scale events (Sánchez Goñi *et al.*, 1999, 2000b).

The western Balkans and, in smaller measure, the Alps and the Italian mountains appear to have been the major broadleaved tree refugia during the last glaciation (Bennett *et al.*, 1991; Zagwijn, 1992; Willis, 1996; Tzedakis *et al.*, 2002), contrasting with the largely sclerophyllous vegetation of the Iberian refugium (Carrión *et al.*, 2000; Figueiral & Terral, 2002). The Near East and south-west Asia, though warmer than the Mediterranean peninsulas, were also more arid and were not, therefore, as important as refuges for temperate plants (Willis, 1996). It is interesting to note, however, that the aridity of Israel was replaced by wetter conditions for much of the period between 40 kyr and the LGM around 20 kyr (Bar-Matthews *et al.*, 1997; Gvirtzman & Wieder, 2001), and that strong north–south climatic gradients existed, as they do today, due to the decreasing influence of the Mediterranean towards the south (Goodfriend, 1999). These crucial differences between each of the major Mediterranean peninsulas and also the Middle East have been overlooked in past considerations of the human occupation of Europe.

The north-west African climate was largely influenced by the southward migration of the dry subtropical high pressure zone during glacials that generated arid conditions (Hooghiemstra *et al.*, 1992; Dupont, 1993). Mediterranean woodland was significantly reduced during glacials at the expense of steppe and semi-desert and regained its importance during interglacials (van Andel & Tzedakis, 1996). The situation in north-east Africa is discussed in the next section in relation to the expansion and contraction of the Sahara.

Africa

The shift towards increased cooling and aridity is detected in Africa after 2.8 Myr and the subsequent pattern of African climate was a continuum of wet and dry conditions (deMenocal, 1995). North-east Africa became progressively more arid with long dry periods interspersed by short pluvial episodes (Crombie *et al.*, 1997). After 200 kyr, African glacial stages were more arid than those of the middle Pleistocene (Jahns *et al.*, 1998).

The complexity of the African climate is the result of the size and heterogeneous nature of the continent. Nevertheless links between Northern Hemisphere climatic conditions and those in tropical Africa are becoming apparent (Johnson *et al.*, 2002). During the late Pleistocene the development of arid conditions and the southward shift of West African vegetation zones were synchronous with the high-latitude glaciations and with correspondingly cold North Atlantic Sea Surface Temperatures (SSTs) – the vegetation responded swiftly to these abrupt changes (deMenocal, 1995; Jahns *et al.*, 1998; Gasse, 2000; Zabel *et al.*, 2001). During cold and arid phases much of Africa between approximately

5° N and 35° N, as well as eastern and southern Africa, saw a flux of shifting vegetation types from Mediterranean sclerophyll woodland (in the continental extremes), through temperate semi-desert, temperate desert, tropical desert, tropical semi-desert, tropical grassland and savanna (Adams & Faure, 1997; Dupont *et al.*, 2000; Salzmann *et al.*, 2002). Corresponding altitude shifts in vegetation occurred in the major mountain blocks (Wooller *et al.*, 2003).

The Sahara changed dramatically during these cycles. During the last interglacial, conditions were much wetter than today. There were significant pluvial episodes, reflected by travertine deposition, in the Western Desert of Egypt (currently one of the driest areas on Earth with a mean annual precipitation of 1 cm) at that time (Crombie *et al.*, 1997). Atmospheric circulation patterns were significantly different from today and the Sahara Desert contracted (Dupont *et al.*, 2000). During the subsequent glacial cycle the desert was even more extensive than it is today (Grove & Warren, 1968; Gaven *et al.*, 1981; Swezey, 2001). Superimposed on these cycles are millennial-scale late Quaternary cycles, reflected in lake level and aeolian sediment deposition fluctuations (Swezey, 2001).

Tropical rainforest and mangrove vegetation correspondingly expanded and contracted in central and west Africa (Lezine *et al.*, 1995) and tropical montane forest responded in similar fashion through changes in elevational distribution (Jahns *et al.*, 1998). In west Africa rainforest and mangrove swamps were widespread during OIS 5 and 1, but largely reduced in OIS 3 and 4 and particularly in OIS 2 and 6 when open, grass-rich, vegetation dominated. The expansion of montane forest during oxygen isotope substages 5d (115–105 kyr) and 5b (95–85 kyr) is indicative of cool events within this interglacial (Dupont *et al.*, 2000) and is probably characteristic of the small-scale global climate variability of the last glacial cycle.

Synthesis

There are several points that will be of particular importance to the discussion of Modern Human colonisation and Neanderthal extinction in the following chapter. In global terms the last glacial cycle was the coolest and most variable of the Pleistocene with wide climatic fluctuations that were particularly bad during glacial maxima. Climate oscillations occurred across a range of temporal scales, including small scale annual to decadal, and transitions were abrupt. The greater part of the cycle was dominated by climatic conditions that were intermediate between those of glacials and interglacials. Sea-levels fluctuated between highstands of +2 to +12 m and lowstands of −118 to −135 m. During the long, intermediate, conditions of OIS 3 sea-levels oscillated around −80 m.

Figure 6.2. Predicted major vegetation transitions in relation to late Pleistocene
climate. Bioclimate boundaries as in Figure 5.3. M-D, Mediterranean
vegetation–desert transitions; M-S, Mediterranean vegetation–steppe transitions; T>,
temperate forest intrusions during warm and wet climate; <S, steppe intrusions during
cold and arid climate; T V, tundra intrusions during cold and arid climate.

The significance of these changes was in the exposure of large areas of continental shelf and not in the opening up of land bridges (the Strait of Gibraltar,
for example, narrowed but remained open throughout). In the Mediterranean,
areas of exposed coastal shelf would have been within the most benign climate
regimes and would have therefore contributed to the surface area of refugia.

Vegetation responses took the form of open vegetation–woodland transitions
across the western Eurasian Plain (Figure 6.2). The greater part of OIS 3 was
treeless from the British Isles to the Russian Plain. Woodland only therefore
covered large areas of the Plain for short periods of the glacial cycle. In the
Mediterranean, where moisture–aridity gradients were particularly critical, the
pattern was more complex on account of topographical heterogeneity with significant local differences in vegetation patterns and responses. Generalisations
are impossible and each area has to be understood independently. This even
applies within the peninsulas, especially in the Iberian Peninsula. OIS 3, in
particular, is complex with moments of expansion of Mediterranean taxa in

some areas and of steppe vegetation expansion in others, adding a spatial component to the climatic unpredictability (Figure 6.2).

North Atlantic sea-surface temperatures played a significant role in modulating climate, and vegetation changes took the form of woodland–steppe transitions, altitudinal movements of plant species, contraction and expansion of Mediterranean thermophyllous species and shifts in the dominance of broad-leaved and coniferous woodland. In the coldest moments, North Atlantic marine mammals and birds reached the Strait of Gibraltar. The Mediterranean peninsulas were pivotal as glacial refugia for Mediterranean and temperate vegetation but the Middle East appears to have been less important on account of long periods of high aridity. There were important differences between refugia, the Italian and Balkan refugia being especially important for temperate broad-leaved trees, and Iberia for Mediterranean taxa. The African coast of the Mediterranean came under the influence of desert and semi-desert during arid periods related to cold events with the expansion of Mediterranean woodland, especially in the north-west, during interglacials. The African picture was dominated by rain–arid cycles with swift vegetation responses. The northward and southward shifts in the belts of desert, grassland and savannah repeatedly opened and closed the door between tropical Africa and the Mediterranean.

7 The Modern Human colonisation and the Neanderthal extinction

There are repeating patterns that we can observe among a wide range of organisms that occupied Pleistocene Europe. These include the contraction into southern refugia and subsequent expansions during climatic amelioration that I shall describe in this chapter. In seeking a generalised theory that accounts for the varying fortunes of the Neanderthals and Moderns we must consider that Pleistocene people were humans, not super-humans. By this I mean that, even though humans in the Pleistocene had succeeded in evolving socio-cultural and technological achievements that undoubtedly set them apart and gave them great advantages over the other animals with which they shared territory, they were by no means independent of the environment that surrounded them and were very much subject to the forces of natural selection. If we are able to see similarities of pattern with other Pleistocene animals then we will have advanced towards a generalised theory. If we are unable to find such similarities then we will also have advanced in our understanding of the distinctness of humans in the Pleistocene world.

Humans, climate and environmental change

Eurasian humans throughout the Pleistocene were restricted to southern refugia during cold episodes. The degree of permanence of human populations would have been highest in tropical and equatorial regions with decreasing probability of permanence away from these areas (Finlayson *et al.*, 2000a). The most significant general pattern is the permanence of many species along the southern part of the European range in the Mediterranean peninsulas of Iberia and the Balkans, in particular, and the temporary and often brief range extensions into northern lands (Hewitt 1999, 2000). Every time the glaciers and ice sheets advanced so populations were confined to the Mediterranean refugia. That humans responded in the same manner as most other organisms is undeniable and it affected Neanderthals and Moderns as it no doubt affected their predecessors. The Neanderthals (including their European ancestors) survived in Europe (Figure 5.3) for over 400 kyr but it is important to note that they:

(1) only occupied areas of the central and western European Plain during milder events;

(2) they never colonised the steppes of eastern Europe;

(3) They were restricted to the Mediterranean peninsulas (and Crimea and the Caucasus) during the colder episodes.

Neanderthals therefore were restricted to southern refugia during cold and arid events and they were unable to recover from the last of these (Finlayson *et al.*, 2000a). Moderns were not much better at dealing with the glaciations. The first major glacial advance that hit them in Europe forced them into the same southern retreats that Neanderthals had entered previously. They managed to hold out, just as populations of Neanderthals had done during earlier cold events, and subsequently they spread north (Torroni *et al.*, 1998, 2001). Humans have not experienced another cold episode since. The observed pattern for Neanderthals is therefore no different from that of other human populations, including the Moderns, and is part of a recurring theme.

In my view this sets the large-scale spatial and temporal scenario that best explains the changes in human populations that occurred in Europe and Asia throughout the Pleistocene. In an earlier paper (Finlayson *et al.*, 2000a) I have indicated the conditions that would have favoured the spread of human populations from tropical Africa into the Middle East and from there towards Asia and Europe (see also Chapter 3). I have also proposed that, once in South-east Asia, human populations would have achieved degrees of permanence comparable to the African populations and such populations would have functioned as secondary sources from which temperate Asian populations were fed (Chapter 3).

The case for continuity in human evolution is strongest in equatorial and tropical areas of the world. Given that the origin of the lineage that led to the Moderns was African (Chapter 4), we should observe the longest period of continuous human occupation in that continent. This should be followed by tropical and equatorial Asia, the difference with Africa being historical. The degree of persistence of human populations away from these areas at any given time would have varied with latitude and altitude. The ability to colonise and persist further and further away from the tropics improved through time. Behavioural mechanisms evolved in the open tropical savannahs that pre-adapted populations for colonisation away from the tropics. I predict therefore that extinction of local and regional human populations was probably a feature of non-tropical areas and that the probability of extinction decreased with time.

The extinction of a human population, such as the Neanderthals, in Eurasia during the Pleistocene would not have been a singular event (Finlayson, 2003). The ultimate causes of human population extinctions in the Pleistocene are probably very similar in all cases. Populations in southern refugia became fragmented and were unable to recover. Climatic conditions, acting on habitats and resources, were primarily responsible for range contractions and population fragmentation and reduction (Finlayson *et al.*, 2000a). The probability of persisting through a bad event would have depended on the intensity of the bad event, the frequency of bad events, the intensity and length of intermediate good events allowing population recovery, initial population size and demographic and genetic population parameters. Socio-cultural and technological attributes may have alleviated situations in some circumstances (Gamble, 1999).

A single proximate cause of local and regional human population extinctions in the Pleistocene is unlikely. The retreating rear edge of a range during a period of contraction will be expected to suffer severe shrinkage, dissection and extinction with a severely bottlenecked last surviving population (Hewitt, 2000). Once human populations became fragmented and depressed to the point of imminent extinction, the final cause of extinction would have varied from one situation to the next. Proximate extinction causes could have included stochastic processes, local inbreeding, competition, habitat and resource fragmentation, Allee effects, disease and reduced immunity (Figure 7.1). It is therefore pointless, given current data resolution, to seek a single proximate solution to explain the extinction of the Neanderthals, or indeed any other human group.

Before changing the subject I want to discuss one final point, and a crucial one, in understanding extinction. This is the effect of frequency as well as intensity of environmental fluctuations, time lags and cumulative effects. A number of authors have sought direct correlations between environmental fluctuations and demographic changes. People have looked at climate curves and attempted to prove or disprove effects on populations by seeking direct matches between the two. These have produced conflicting results when trying to interpret the effect of climate on human population dynamics and especially the Neanderthal extinction (e.g. Courty & Vallverdu, 2001). The absence of a correlation need not, for example, be evidence for no environmental effect. Environmental effects may be expressed in many ways and at different scales. In the case of the Neanderthal extinction we are looking at a large scale effect that depresses populations globally and the effect is caused by an increase in frequency of climatic oscillations, i.e. increasing instability. Why did the Neanderthals not become extinct earlier during a similar period? This presupposes that, for example, starting population sizes were equivalent before each perturbation. Put simply, the effect of 50% population reduction in a population of 10 000 may allow recovery but the same effect on a population of 100 may well lead to extinction. Theory predicts that in the case of two species with different colonisation (c) and extinction (e) rates but equal c/e values, the species with higher c and e values will reach a new equilibrium after habitat destruction faster than

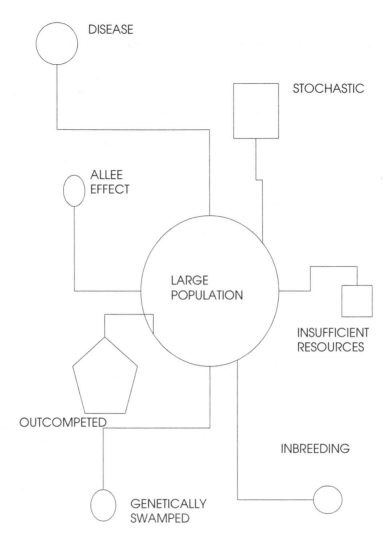

Figure 7.1. Potential causes of extinction of local populations fragmented from a hypothetical large parent population.

one with lower *c* and *e*. This is an example of relaxation in which the new equilibrium level of patch occupancy is not reached instantly. We may say that the species exists as 'living dead' (Gilpin & Soulé, 1986; Groom & Pascual, 1998). So matching the moment of extinction with an environmental event at that moment would be absurd! The irony is that the species might actually become extinct during favourable climatic conditions! There is a practical point that we must also consider. That last Neanderthal populations on record occur around 31–28 kyr. Trying to match precise climatic conditions to these dates is not only unrealistic, because of what I have said so far, but also because these are not real extinction dates. These are dates when populations were still high enough for us to detect them in the archaeological record. So, as with other things we have looked at so far, we can only look at the Neanderthal extinction from a large-scale perspective because we simply do not have the resolution to go further. Some people may persist in trying to find the cause of the death of 'the last of the Neanderthals'. It is like looking for the missing link. I prefer to stay with the view that high environmental instability depressed and fragmented their populations at the end of Oxygen Isotope Stage (OIS) 3 beyond recovery. Such a view has theoretical and empirical support. If the rate of environmental movement is slow, species will be expected to track their particular environments across space as geographical range changes are more malleable than morphology or environmental tolerance (Pease *et al.*, 1989). The Neanderthals appear to have tracked their environments in this way. When the rate of change intensified towards the end of OIS 3, they went extinct.

Competition

Competition structures communities that are in equilibrium and is not important in situations of wide environmental fluctuations and unpredictable disturbance (Wang *et al.*, 2002). Finlayson *et al.* (2000a) have clarified the situations in which ecological competition was likely to have occurred in Late Pleistocene Europe and western Asia and came to the conclusion that, if it ever took place at all, competition between Neanderthals and Moderns would have been ephemeral and would not have determined the final outcome of the two populations. Similar situations would have arisen in other parts of the world. Rolland (1998) and Richards *et al.* (1998) comment on the sparse, low-density, population pattern for Eurasia in the Pleistocene, suggesting that demographic carrying capacity was not attained, and have contrasted this with the situation in Africa. Harpending *et al.* (1998) estimated the effective human population size not to have exceeded 10 000 for most of the Pleistocene.

Van Peer (1998) found two coexisting (archaic and modern) populations in north-east Africa in the late Pleistocene. One (archaic) was exclusively riverine-adapted and only occasionally used desert. The other (modern) allowed populations to adapt to varied environments, including the desert. Occasionally, depending on prevailing conditions, the two systems functioned in the same area.

The situation was probably similar in the Middle East where Moderns and Neanderthals would have been within the same geographical area for over 60 thousand years (Tchernov, 1992, 1998; Bar-Yosef, 1994, 1998) and in other zones of heterogeneous landscapes such as along the edge of the Russian Plain (Soffer, 1994). Away from these contact zones one or other form would have been distributed over large areas with minimal contact. These data are in keeping with theoretical predictions that show that environmental heterogeneity effectively supports long-term coexistence of very similar species (Hanski, 1983; Taneyhill, 2000).

A case within recorded history illustrates how two human forms (contemporary *sapiens*) coexisted in a geographical area and how one became extinct subsequently as the direct result of climate change. The work in Greenland (Barlow *et al.*, 1998; Buckland *et al.*, 1998) recreates a scenario that indicates that only one of two (Norsemen and Inuits) existing modern human groups survived the medieval 'Little Ice Age'. Although the Norsemen had been the earlier colonisers and were apparently well-adapted to their environment, they failed to survive a period of extended cold. In contrast, however, available evidence points to there being sufficient, if not abundant, resources for the Inuits at a time when the Viking settlers were having to slaughter their animals for food. This points to significant lifestyle differences between the farming, relatively sedentary, Norsemen and the nomadic and wide-ranging Inuits as being one, if not the main, factor in determining the survival of these groups. There is no evidence of direct competition nor is it suggested as a possible factor in the 'extinction' of the Norsemen. These two forms behaved as ecotypes and the analogy with the Modern–Neanderthal situation in the late Pleistocene is evident. There are other excellent examples that show the effects of climatic and environmental changes on recent human populations (Binford *et al.*, 1997; Cachel, 1997; Park, 1998; Sterling, 1999; Verschuren *et al.*, 2000; deMenocal, 2001; Anderson, 2002; Axtell *et al.*, 2002; Weber *et al.*, 2002; Weiss, 2002) that serve to illustrate that population expansions, crashes and extinctions have continued to occur in humans independently of competitive processes even into historical times.

A popular thesis is that competition from the newly arrived and 'superior' Moderns caused the extinction of the Neanderthals. The only basis for the argument is that of an apparent association between the arrival of the Moderns into Europe and the extinction of the Neanderthals. Inter-specific (or inter-population) competition is a very difficult phenomenon to demonstrate in any extant wild population today. For the conditions of competition to apply the competing populations must be at environmental carrying capacity and must use similar resources and obtain them at the same times and in a similar enough manner to make one population's rate of increase have an effect on the other's. If the populations are not at carrying capacity then a situation of resource superabundance exists and there is no competition.

It is practically impossible to know whether or not Neanderthals and Moderns were in competition. I suggested, in Chapter 5, that the two forms differed in landscape use. This would explain how, in the Middle East for example, Neanderthals and Moderns could have co-existed within the same geographical area over protracted periods. Furthermore, the variability of resources as a result of climatic oscillations is unlikely to have led to a stable situation that permitted the two forms to reach carrying capacity and equilibrium populations within the same geographical area. Coexistence therefore was no more than fluctuating levels of two populations below carrying capacity, each with a particular mechanism of use of space evolved in different landscapes and geographical areas.

In terms of competitive ability, one could suggest that in a situation of an expanding population of Moderns and a stable (or even locally growing during cool periods when the range boundary shifted southwards) population of Neanderthals in the Middle East, the likelihood would be that the resident population (in this case the Neanderthals) ought to have a competitive edge over the pioneers. Knowledge of the territory, its resources and mechanisms of optimal resource collection would weigh heavily in their favour. An expanding population of pioneers would only succeed if it behaved in a different manner and the conditions favourable for the resident changed. This is what seems to have occurred in the Middle East. For a long time the two forms occupied the same geographical area. For much of this time the Neanderthals were probably on well-established territories and the Moderns would have occupied areas marginal to the Neanderthals. This would have been particularly likely in areas like the Middle East with a heterogeneous mosaic of habitats from mountains to lowland plains and desert. Wang *et al.* (2000) have shown that two ecologically identical species can coexist when there is a trade-off between local competitive ability and invasion ability. If we take the view that I take in this book that the Neanderthals were probably locally competitively superior to Moderns, but Moderns had greater dispersal capacities, then we have here a theoretical basis for long periods of geographical coexistence in spite of ecological similarity (Chapter 5). There would have been a shifting balance between the two populations, a kind of semi-permanent geographical coexistence. The population of Moderns would have expanded when its favoured habitats expanded and its range slowly penetrated the mountains to the north that acted as a physical barrier to dispersal. On reaching the plains of Eurasia the limits on this population were removed and there was a rapid ecological release and expansion. The colonisation of the Eurasian Plain, free from physical barriers, rich in resources and largely free of Neanderthals had to be rapid, and it was.

If the Moderns arriving in Europe from the east had been responsible for the Neanderthal extinction, then we would expect an east–west pattern of extinction as the Moderns arrived. Instead we observe a pattern of extinction that is related to bioclimatic zones strongly suggesting that a climate-driven impact was responsible (Figure 7.2). I do not, therefore, consider competition from Moderns

Figure 7.2. Location of late (N, post-35 kyr) Neanderthal sites in relation to bioclimate. Grey, Mediterranean. Bioclimates after Rivas-Martínez (1996). Bioclimate boundaries as in Figure 5.3.

to have been a significant cause of the extinction of the Neanderthals. This is probably also so in the case of other archaic forms. If there was competition, it is likely to have been highly localised and would in all likelihood have favoured the well-established local populations. Colonisers would only have succeeded in situations where local groups were in a phase of recession, such as occurred in south-west France around 40–35 kyr.

Hybridisation

An ecotype is a genetically distinct form that occurs in a specific habitat but which interbreeds more or less freely with another form that occurs in an adjacent habitat (Cain, 1971). There are many intermediate forms in nature between ecotypes and good species that exclude each other in different habitats but with contiguity and no hybridisation. In some cases the observed hybridisation is secondary (Mayr, 1963; Cain, 1971), that is when two forms that differ significantly in their genetic makeup meet comparatively recently in the wild and a range of hybrids is possible. Cain (1971) considers that most geographical subspecies should probably be considered ecotypes with a single area of occurrence and I consider that this terminology is one that is appropriate to the Modern Human/Neanderthal situation in areas of recent contact.

When hybridisation is too small to be significant in determining species limits and when it is so high that the hybridising forms should be regarded as having combined to form a new species is unclear (Cain, 1971). In particularly heterogeneous geographical areas such as Iberia, a mix of levels of interaction is far more likely (see Transition below). Regions of high overlap would occupy only a small proportion of the geographical area. On current evidence I do not consider that there was any significant, large-scale, hybridisation between Moderns and Neanderthals.

When populations expanding from glacial refugia met they often formed hybrid zones (Barton & Hewitt, 1985; Hewitt, 1996, 1999, 2000). The main ones in Europe are centred in the Alps and central Europe as well as the northern Balkans and the Pyrenees. Such boundaries tend to be narrow, tension, zones and are marked by a reduction of hybrid fitness, the extent of which determines the zones' width. Until climate changes a situation some hybrid zones may become 'trapped' in local areas of low density or dispersal. These hybrid zones may additionally act to protect the integrity of the genomes on either side until a subsequent glaciation reduces the two to separate refugia (Hewitt, 1996). In the case of the expanding population of Moderns and the receding population of Neanderthals in late OIS 3 Europe we may expect that, given the rapidity of climate change, any existing hybrid zones between the two forms would not have been stable and would either move, in this case in a north-east to south-west manner, or disintegrate (e.g. Carney *et al.*, 2000). Recent work has shown that such movement of hybrid zones, in response to environmental perturbation (Bynum, 2002), may be important in biogeography and evolution (Dasmahapatra *et al.*, 2002). Finally, we should not overlook the possibility of hybrid zones existing between Neanderthal populations as these recolonised areas of the Eurasian Plain from Iberian and Balkan refugia.

Human populations would have been repeatedly isolated from each other (Finlayson, 2003). The surviving populations would have re-met during periods of population expansion. The degree of inter-breeding would have varied from total inter-mixing to complete isolation. The probability of two human populations inter-breeding when coming into contact would have depended on the degree of genetic distinctness of each population which, in turn, would have been dependent on the degree of prior isolation. The question of inter-breeding of previously isolated populations would not just have affected the Moderns and Neanderthals, which is the case that is receiving greatest attention today

(Chapter 4; Duarte *et al.*, 1999; Tattersall & Schwarz, 1999), but different Neanderthal populations that had been isolated in different refugia. Similarly the degree of inter-breeding between Moderns and Neanderthals may have varied between regional and local populations of each. On current genetic evidence we can conclude that there was no long-term Neanderthal genetic contribution to the present-day human gene pool (Chapter 4). It is impossible, on present evidence, to assess the degree of inter-breeding and the contribution of Neanderthal and early Upper Palaeolithic Modern genes to each other's populations.

I therefore predict that human range expansions and contractions were frequent and of varying extent, only the most widespread and intense being recorded in the fossil and archaeological record. There would have been spatio-temporal variability in the degree of secondary hybridisation and contact between populations that became allopatric during glacial events (Cain, 1971). Behavioural, morphological and physiological isolating mechanisms, when present, would have acted to maintain population distinctness.

Behavioural differences and cultural exchange

The cognitive abilities of the various populations of the *sapiens* polytypic species (that included Neanderthals and Moderns) would appear to have a common and distant origin and the taxonomic definition of human populations is arbitrary (Chapter 4). The human lineage may be regarded as a single palaeospecies with geographical populations of varying degrees of distinctiveness at any particular point in time. Thus for any time period it should be productive to consider global human populations as forming a polytypic species complex (Chapter 4). Evidence from Africa in particular indicates that behaviour attributed to 'modernity' as part of the European Upper Palaeolithic Revolution has much earlier origins (Chapter 5). Behaviour, including social behaviour, culture and technology, would have evolved as adaptive responses to specific situations. Responses to similar environmental and social pressures would have been met by similar, though not necessarily the same, solutions. Cases, such as the development of so-called 'Upper Palaeolithic' or 'transitional' technologies, should be seen from the perspective of behavioural convergence. This should nevertheless not negate the possibility of cultural and technological information exchange where different groups met and interacted. As with other biological aspects dealt with in this book, a mosaic of possibilities would have existed and it is unrealistic to seek a common solution.

The question of acculturation or the independent evolution of cultural attributes is of considerable controversy today, particularly in the context of the arrival of the Aurignacian in Europe and the emergence of Middle Palaeolithic

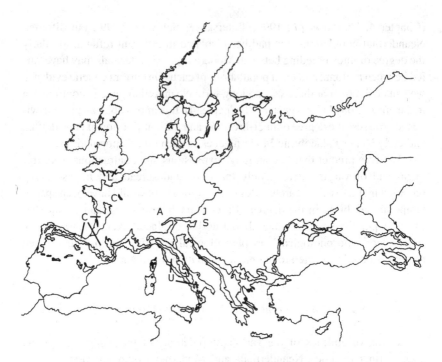

Figure 7.3. Distribution of 'transitional' industries in relation to bioclimate.
C, Chatelperronian; U, Uluzzian; A, Altmuhlian; S, Szeletian; J, Jerzmanian;
L, Lincombian. Bioclimates after Rivas-Martínez (1996). Bioclimate boundaries as in
Figure 5.3. Transitional industries after Raposo (2000).

technologies with Upper Palaeolithic elements (d'Errico *et al.*, 1998; Mellars, 1999; Zilhao & d'Errico, 1999). I view this as a problem of contact and I am of the view that contact in the case of Moderns and Neanderthals in Europe was restricted on account of low population sizes and habitat differences. We have also seen the degree of fluidity in the adoption of Middle or Upper Palaeolithic (or Middle Stone Age/Late Stone Age) technology in response to environmental change, both in Moderns and Neanderthals (Chapter 5). In a cooling world in which the Neanderthal populations were being forced into refugia, it could be expected that technological changes that favoured existence in marginal zones would be favoured. The presence of the intermediate technologies occurs precisely in such intermediate areas between the heterogeneous zones of Europe and the plains and clearly they must be adaptations to a changing resource and habitat structure world (Figure 7.3). Given the degree of flexibility between the types of technology adopted by Moderns and Archaics across the world it is not unreasonable to expect behavioural responses such as those documented in Europe to evolve repeatedly in isolated populations. For example, in the late Middle Palaeolithic population of the Nile there were two populations: those with a new material culture from the south; and those that developed a local complex that evolved and was therefore not replaced by technological change. In north-east Africa the transition was independent in the two groups and after a long period of coexistence. It may have happened in the context of enhanced social interaction and probably because of it (Van Peer, 1998). The degree of coincidence with the arrival of Moderns does leave the door open in respect of whether such behaviours arose independently or whether, instead, they were obtained by copying (acculturation). In north-east Asia the early presence of the Upper Palaeolithic, around 42 kyr, is characterised by blade production, with the retention of Levallois cores, but the retention of Mousterian technology as late as 35–28 kyr (Brantingham *et al.*, 2001) is a further indication of the functionally adaptive nature of these industries and of the duration of the transition, especially in zones of contact between the plains and the heterogeneous belt. Interestingly, early Upper Palaeolithic blade technology is associated with open air sites but its geographical distribution remains largely in the heterogeneous environments south of 55° N. It is with the opening of areas to the north, that were covered by boreal forest and were replaced by mammoth-steppe, after 30 kyr that we observe the typical Modern Human pattern of plains exploitation (Goebel, 1999). These examples illustrate the complex interactions between Moderns, Neanderthals and their adaptation to changing environments. In ecotonal areas where the plains met the mid-latitude belt, such as in southern Siberia, we observe the degree of experimenting that led to the invention of new ways of exploiting the changing environments.

The nature of the landscape would have been largely responsible, at any stage, for the adaptive behavioural ecology characteristics of each human population. It is expected that human populations selected regions, landscapes and habitats that maximised their fitness. Areas high in biomass or in diversity would have been the prime focus. In the first case the open savannahs of Africa and the great plains of Eurasia would have been particularly favourable. In the second case ecotones, zones with a number of distinct habitats over a small area, would have been optimal. Ecotones would have included coastal areas, lake margins, riverside habitats and topographically heterogeneous zones. In the latter case the mid latitude belt from Portugal and the Maghreb to the Caucasus and the Altai would have presented a large mass of heterogeneous landscape.

The plains of Eurasia would have been, always in the east and during cold/arid events in the centre and west too, homogeneous in human terms. In the west the development of forest, forest margins and the presence of lakes during mild

oceanic phases would have opened up opportunities for ecotonal human ecotypes and reduced them for plains human ecotypes. Morphology and behaviour
would have been major components permitting different human populations
to successfully exploit plains or ecotones. In this book I have suggested that
the long-limbed, gracile, morphology of Moderns, coupled with an appropriate
social and behavioural lifestyle, suited them particularly well to the long-range,
highly mobile, system of the plains (chapters 4 and 5; Finlayson *et al.*, 2000a).
Similarly, the more robust morphology of the Neanderthals would have been
less suited for an open plains existence and the evidence of severe limb wear
would appear to confirm this view. Thus, Neanderthals living in the ecotonal
conditions of the heterogeneous landscapes of southern Eurasia and used to exploiting a range of resources over a small area, would have extended northwards
into the Eurasian Plain when mild conditions induced the spread of the forests
and generated an extension of the ecotonal conditions. It is not surprising, therefore, that Neanderthals never colonised the steppe environments of the eastern
European Plain even though they lived close by in the hills and mountains of
Crimea, the Caucasus and the Altai (Soffer, 1994). Similarly, when cold and
arid conditions took hold the range of the Neanderthals receded as the woodland of western and central Europe became steppe. It is in edge areas that we
would expect the greatest stress as populations attempted to adapt to the rapidly
changing landscape. These areas would have included south-western France,
the Italian Peninsula, the northern and central Balkans, hilly landscapes in central and eastern Europe and sites along the edge of the Russian Plain. It is in
these areas that we would predict the presence of 'Upper Palaeolithic' technologies among Neanderthal groups as they attempted to adapt their tool kit to
the changing circumstances and in the direction of the plains dwellers that were
used to exploiting such environments (Figure 7.3). In such a scenario I would
also predict that the last Neanderthals would have lived close to the topographically heterogeneous zones. Within these, the populations in edge zones would
have attempted to adapt technologically whereas those in core areas (such as
Iberia) would have maintained a traditional technology to the end.

If my interpretation is correct, then the Neanderthals were a people of mid-
latitude Europe that were able to extend their geographical range northwards
during mild events. They evolved in the rich heterogeneous landscapes of mid-
latitude Europe and their morphology was best suited for the kind of rugged terrain and close-quarter hunting that the landscape demanded (chapters 4 and 5).
As with many other animals, attributes of exploitation of such landscapes would
have included small home ranges, diverse diet at the annual scale as different
resources were seasonally cropped, small population units that were in constant
contact as they moved across the home range, precocious children that would
be able to move with the adults at an early age and an intimate knowledge of the

home range and the seasons (Chapter 5). The down side of such a strategy would have been increased likelihood of fragmentation and isolation with consequent genetic effects.

The Moderns most probably entered the Eurasian Plain somewhere in the region between the Black and the Caspian Seas. Whichever way it was, by 40 kyr we see the expansion of the geographical range of this form. The rapid expansion shows the characteristics of an ecological release and the flat landscapes of the Eurasian Plain undoubtedly played a catalytic role as they would persistently throughout history (for example for the huns or the avars). The nature and distribution of resources determines home-range size. The exploitation of the plains required large home ranges and a distinct social system and probably a greater within-group division of labour, centralised base camps and systems of storage that would only be possible if such bases existed (Chapter 5). Life in the plains would have been demanding, not least being the reduction in winter daylength and the great reduction in resource range compared to tropical and mid-latitude areas. As group components were separated for periods of time, there would have been a greater pressure for the development of sophisticated communication and social-binding systems so it is perhaps not too surprising to find so much art and ornamentation in these groups.

In the final analysis there is therefore very little difference between Neanderthals and Moderns. They exploited the same range of food resources and had similar technological abilities. Observed differences reflect population adaptation and there are no linear, directional, trends. There is no clear Modern–Neanderthal boundary that cannot be explained by differences in ecological setting. Moderns differed from Neanderthals in adaptations (morphological and behavioural) that enabled them to operate at larger spatial scales. The high frequency of climatic oscillations and the trend towards cooling towards the end of OIS 3 introduced environmental instability (Chapter 6). The exploitation of heterogeneous landscapes, as we have seen, was the Neanderthal way of dealing with short-term instability. Prolonged instability meant that the scale of Neanderthal response did not match the scale of the perturbations. The Moderns, on the other hand, could deal with such large-scale instability because they operated on larger scales. The expansion of favoured open, homogeneous, landscapes and their associated faunas, further enhanced their probability of survival.

Glacial refugia

The inescapable consequence of the climatic fluctuations of the Pleistocene for many animals and plants were the huge changes in geographical distribution

(Hewitt, 1996, 1999, 2000). Some species were able to maintain themselves in southern European refugia for a number of glaciations while others have arrived more recently. Extinction would have been a feature of the glaciations even in southern refugia (Hewitt, 2000; O'Regan *et al.*, 2002). O'Reagan *et al.* (2002) have highlighted the importance of southern refugia in the extinction process of large carnivores, with chance playing a major role in the survival of the reduced and isolated populations. Such was the case of the Neanderthals, probably originating from a recent European arrival (*c.* 500 kyr) and managing to survive several glaciations in southern refugia before finally becoming extinct just before the Last Glacial Maximum (LGM).

An increasing number of studies are clarifying the generalised responses of European populations of many organisms to these climatic fluctuations. Taberlet *et al.* (1998) and Hewitt (1999, 2000) have summarised the patterns. The Balkan Peninsula was a refuge that acted as the source for recolonisation by all species in the east and also many in the west. Turkey and the Black Sea–Caspian Sea region also appear as refugia. Italian populations, on the other hand, rarely repopulated Europe, the Alps apparently acting as a significant barrier. The Pyrenees were also a barrier to populations dispersing from Iberia but it seems that they were more porous than the Alps. Finally, there is evidence of isolated patches further north, along the southern edge of the steppe–tundra zone, which acted as local refugia (Willis *et al.*, 2000, 2001; Carcaillet & Vernet, 2001; Stewart & Lister, 2001). I suggest that the Balkans refugium, always being more continental in characteristics than the oceanic Iberian refugium, may have additionally held populations that were physiologically better able to expand into temperate areas in the initial stages of a deglaciation. This may explain the importance of this refugium for temperate trees (Chapter 6; Bennett *et al.*, 1991).

The Iberian refugium

In this section and the next I use the Iberian Peninsula as a model for the study of human dispersion and dispersal during the Quaternary. Iberia is diverse and large enough to act as a natural laboratory for the study of human interactions in the Pleistocene. The southern Iberian Peninsula has been occupied by humans since at least 500 kyr but probably significantly earlier. I proposed in Chapter 3 that the hominids that were the ancestors of those inhabiting Atapuerca over 780 kyr (Bermudez de Castro *et al.*, 1997) may have reached Europe across the Strait of Gibraltar and there is also a claim of hominid occupation in Orce (Granada) at 1.2 Myr (Oms *et al.*, 2000) which must await further evidence. The southern Iberian Peninsula has been a crucial region throughout this period, acting as a refugium for human populations during glaciations (Finlayson, 1999; Finlayson *et al.*, 2000a; Straus, 2000), being one of the areas of late Neanderthal survival (Vega-Toscano, 1990; Finlayson, 1999). Giles Pacheco *et al.* (2003) examined the distribution of humans in southern Iberia (Andalucía and Gibraltar) after 500 kyr based on an inventory of archaeological and palaeontological sites. They analysed these data against climate for the period 90–0 kyr (GRIP, 1993) at the scale of 0.5 kyr to test the relationships between climate parameters and human distribution.

Giles Pacheco *et al.* (2003) surveyed the literature to identify sites that belonged to distinct archaeological periods (hereafter referred to as cultures) in southern Iberia. The following divisions were established.

Late Acheulian (Mode 2/3)
The Acheulian was established in Europe by 500 kyr (Foley & Lahr, 1997). The data used by Giles Pacheco *et al.* (2003) represented the late Acheulian which is characterised by the standardisation of the use of flint and a generalised introduction of Mode 3 (Giles Pacheco *et al.*, 1993, 2003), and was represented by sites leading up to the last interglacial.

Mousterian (Mode 3)
The Mousterian appeared in Europe by 250 kyr (Foley & Lahr, 1997). It is characterised by the use of the Levallois method of extraction (Klein, 1999) and by a homogenisation of the use of flint and the standardisation of types. It was represented by sites that date from before the last interglacial to 31 kyr (Finlayson & Giles Pacheco, 2000).

Aurignacian (Mode 4)
The Aurignacian, generally associated with Modern Humans, appeared in Europe around 45 kyr (Bar-Yosef, 2000). It reached northern Spain by 40 kyr (Straus & Winegardner, 2000) and is very rare in the south to the point that Finlayson *et al.* (2000b) have questioned its significance there. The use of bone, the manufacture of blades and the appearance of parietal art are characteristics of the Aurignacian (Klein, 1999).

Gravettian (Mode 4)
The Gravettian is found in Iberia from 29 kyr (Marks, 2000). It is characterised by the presence of backed elements, abundant burins and the absence of Aurignacian-type thick endscrapers, Dufour bladelets or bone points (Straus & Winegardner, 2000).

Solutrean (Mode 4)
The Solutrean in Iberia spans the period 20.5–16.5 kyr (Straus & Winegardner, 2000). The technology is distinctive with bifacial techniques with concave base and rhomboidal forms, the appearance of peduncular points, an increase in bone technology (Aura Tortosa, 1989; Ripoll López & Cacho Quesada, 1990; Villaverde & Fullola, 1990) and an explosion of parietal art (Fortea Pérez, 1978).

Magdalenian (Mode 4)
The Magdalenian, which spans the period 16.5–11 kyr, is highly diverse and includes bone implements, a reduction in tool size and the appearance of portable art. Parietal art reaches its peak (Aura, 1989; Straus & Winegardner, 2000).

Epipalaeolithic (Mode 5)
The Epipalaeolithic commences around 11–9 kyr (Straus & Winegardner, 2000) and the last populations are indentifiable to around 6.3 kyr (Oliver & Juan-Cabanilles, 2000). The characteristic innovation is the geometric microlith (Fortea, 1973).

Early and Middle Neolithic
The first two Neolithic divisions were considered by Giles Pacheco *et al.* (2003). The Neolithic reached Iberia around 6.5 kyr or 5.4 Cal BC (Zilhao, 2001). It marked the first presence of ceramic with cardial patterning. The Middle Neolithic, with epicardial ceramics, commenced around 5.7 kyr or 4.5 Cal BC (Oliver & Juan-Cabanilles, 2000).

The number of sites within each technological period and time frame was converted to site density by dividing the number of sites by the time span of the technology and multiplying by 1000, thus representing them as sites/millennium (Straus & Winegardner, 2000). Climate data used were for the period 90–0 kyr (GRIP, 1993) at intervals of 0.5 kyr. The parameters used were: mean $\delta^{18}O$ ($^0/_{oo}$) that is an indicator of temperature; and the coefficient of variation (Sokal & Rohlf, 1981) of $\delta^{18}O$. For the analysis of mean $\delta^{18}O$ and coefficient of variation of $\delta^{18}O$, 100 randomly selected samples of $n = 5$ from each period were iterated. By bootstrapping Giles Pacheco *et al.* (2003) attempted to remove sampling effects related to the difference in duration of each cultural period.

Site density increased from the Acheulian to the Neolithic, with the main increase in the Holocene, peaking in the Middle Neolithic, but with a notable increase also during the Solutrean which was significantly higher than predicted by the model (Figure 7.4). The duration of each culture decreased through time and was especially evident in the Upper Palaeolithic (Figure 7.5). Temperature (mean $\delta^{18}O/0.5$ kyr) decreased gradually prior to 20 kyr and then increased after the LGM (Figure 7.6a). Climate variability (coefficient of variation) decreased

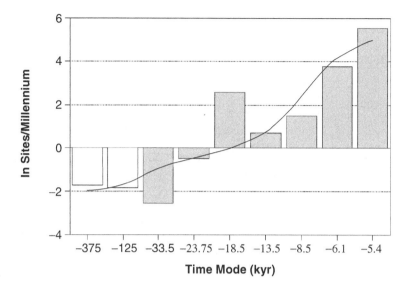

Figure 7.4. Change in density of sites (log n sites/millennium) through time (bars). Curve shows best model fit. The relationship is highly statistically significant ($R^2 = 0.825$; $P<0.001$) and is best described by an S regression model ($\ln(y) = -2.0279 + (-37.975/x)$). White bars, Middle Palaeolithic (Acheulian, Mousterian); grey bars, Upper Palaeolithic (Aurignacian, Gravettian, Solutrean, Magdalenian, Epipalaeolithic, early Neolithic, advanced Neolithic). After Giles Pacheco *et al.* (2003).

through time, especially after the LGM (Figure 7.6b). There was a significant increase in site density with an increase in temperature (Figure 7.7a) and an even stronger relationship with climate stability (Figure 7.7b).

Several patterns emerged from the results of Giles Pacheco *et al.*'s (2003) analysis. There was a trend for site density to increase from the Acheulian to the Neolithic and the rate of increase was greater in the Holocene starting in the Epipalaeolithic (Figure 7.4). There was also a peak during the Solutrean, that had previously been recorded regionally (Finlayson & Giles Pacheco, 2000) and in other parts of Iberia (Straus & Winegardener, 2000). Giles Pacheco *et al.* (2003) interpreted these results as follows: conditions in southern Iberia during the Acheulian and Mousterian and the methods that humans used for exploiting the landscape were such that, at the scale observed, there was very little change during this period. Assuming that site density was in some way proportional to population density, Giles Pacheco *et al.* (2003) concluded that human populations during this long period were constant and low. There was no real change during the first phases of the Upper Palaeolithic. In fact the data suggested a population decline in the Aurignacian and only a slight recovery in the Gravettian

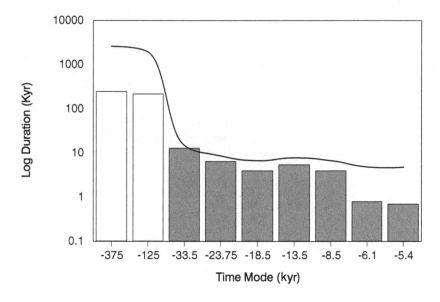

Figure 7.5. Decrease in duration (log duration) of cultural periods through time (bars). Note the significant inflexions at the start of the Upper Palaeolithic and the Neolithic. Curve shows best model fit. The relationship is highly statistically significant ($R^2 = 1.000$; $P < 0.0001$) and is best described by a cubic regression model ($y = 4.4353 + 0.469x + 0.0248x^2 + 5.8 \times 10^{-5}x^3$). White bars, Middle Paleolithic (Acheulian, Mousterian); grey bars, Upper Paleolithic (Aurignacian, Gravettian, Magdalenian, Epipalaeolithic, early Neolithic, advanced Neolithic). After Giles Pacheco *et al.* (2003).

(Figure 7.4). These results are in keeping with the view (see next section) that there was a hiatus in southern Iberia between the extinction of the Neanderthals and the arrival of the first modern humans. The dramatic increase during the Solutrean was considered too great to be a mere artefact of sampling. Its coincidence with the LGM was in keeping with the thesis that it was a phenomenon that reflected a 'refugium effect' (Straus, 2000) at a time when humans were virtually confined to the southern European peninsulas (Gamble, 1999). The results were indicative of populations that were tracking steppe environments, a phenomenon that commenced with the Aurignacian in the central Eurasian Plain (Otte, 1994; Semino *et al.*, 2000; Finlayson, 2003). This conclusion was supported by the apparent population decline during the Magdalenian (Figure 7.4) that suggested that these populations continued to be adapted to steppe environments that were receding at the expense of forest with the post-LGM warming (see also Chapter 8). It coincided with a density increase in northern Spain and in sites at higher elevations (Straus & Winegardner, 2000) at this time which was in keeping with an 'inverse' resource tracking and was

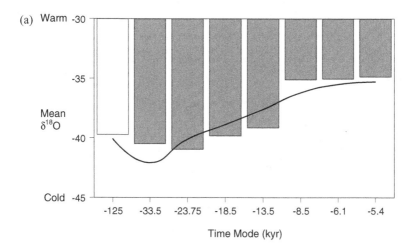

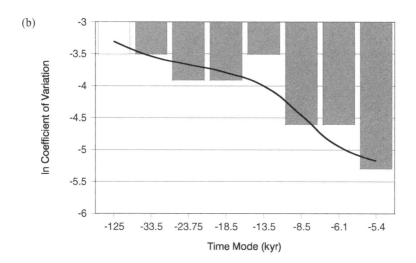

Figure 7.6. (a) Pattern of temperature by time periods related to major cultural periods (bars). Curve shows best model fit. The relationship is highly statistically significant ($R^2 = 0.836$; $P = 0.011$) and is best described by a quadratic regression model ($y = -33.583 + 0.3271x + 0.0022x^2$). After Giles Pacheco *et al.* (2003). (b) Pattern of climatic stability (log n coefficient of variation) by time periods related to major cultural periods (bars). Curve shows best model fit. The relationship is highly statistically significant ($R^2 = 0.806$; $P < 0.002$) and is best described by an *S* regression model ($\ln(y) = -3.2242 + (-10.5197/x)$). After Giles Pachecho *et al.* (2003). For explanation of bars see Figure 7.5.

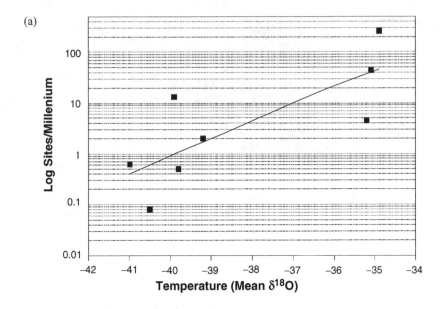

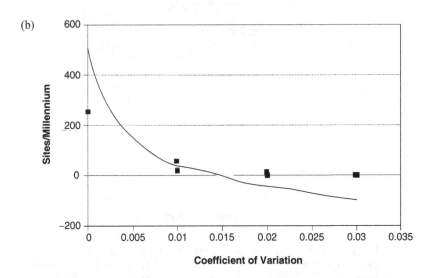

Figure 7.7. (a) Relationship between density of sites (log sites/millennium) and temperature. Curve shows best model fit. $R^2 = 0.604$; $P = 0.023$. The relationship is best described by an exponential regression model ($y = 2.0 \times 10^{13}$ ($e^2(0.769x)$)). After Giles Pacheco *et al.* (2003). (b) Relationship between density of sites and climate stability. Curve shows best model fit. The relationship between site density and climate stability is even stronger than in (a) ($R^2 = 0.95$; $P < 0.005$) and is best described by a cubic regression model ($y = 503.737 - 78268x + 3737111x^2 - 6.0 \times 10^{-7}x^3$).

consistent with genetic evidence of south-west to north-east post-glacial dispersions (Torroni *et al.*, 1998, 2001). The population recovery that commenced at the beginning of the Holocene preceded the Neolithic (Figure 7.4) and suggests an adjustment of local hunter–gatherer groups to the stabilising climatic conditions. The massive and unprecedented subsequent Neolithic population increase reflected a combination of rapid colonisation by eastern populations (Semino *et al.*, 2000; Zilhao, 2001) and an increase in environmental carrying capacity through the introduction of production economies aided by warm and stable climatic conditions.

The duration of cultures decreased significantly and the rate of decrease accelerated with the onset of the Upper Palaeolithic (Figure 7.5). This indicated a significant quantitative change with the arrival of modern humans at a scale that was only subsequently matched with the onset of Holocene conditions and the introduction of the Neolithic (Figure 7.5). This significant shortening of cultural periods reflected an increase in cultural turnover. The Upper Palaeolithic shift may reflect, in some measure, increased mobility and enhanced social networks (Gamble, 1986, 1999; Finlayson *et al.*, 2000a), and therefore an increase in cultural interchange, in people with morphologies very suited to such a landscape exploitation strategy. The Neolithic change may reflect the increased turnover related to population migrations (Zilhao, 2001).

The climatic trends during the period studied were of increasing temperature and climatic stability through time (Figures 7.6a,b). The trends accelerated significantly after the LGM. Site density increased with temperature and climatic stability (Figures 7.7a,b). A very significant result of this study was the much stronger relationship with climatic stability than with temperature. This result supports the view (Finlayson *et al.*, 2000a; Finlayson, 2003; this book) that climatic instability was a major factor in the distribution and abundance of human populations during the Pleistocene. Finlayson & Giles Pacheco (2000) have shown that the distribution pattern of sites of human occupation in the late Pleistocene in southern Iberia shifted from use of open air sites to an increasing use of cave sites, especially in the Upper Palaeolithic. The relationship reported by Giles Pacheco *et al.* (2003) between site density and temperature and climate stability indicates that there have been significant human responses to climate change that have included demographic and dispersion responses. As Finlayson & Giles Pacheco (2000) and Finlayson *et al.* (2000a) have indicated this has meant that there may have been times of climatic instability that effectively generated a depopulation of southern Iberia, an effect that has also been reported for Italy and the Balkans (Raposo, 2000), Central Asia (Davis & Ranov, 1999) and the Middle East (Bar-Yosef, 1996) at the same time. The most conspicuous case is the population response to the climatic instability of

OIS 3 that led to the Neanderthal extinction (see next section). There are two events in the period studied that depart notably from the modelled trend. The first is the Solutrean expansion that coincided with the LGM and the second was the Neolithic population expansion. The latter is well-documented to relate to the arrival of populations from outside the region (Zilhao, 2001). The Solutrean demographic expansion is also likely to be at least in part a reflection of the arrival of humans, probably steppe-adapted from the north, as steppe environments closed in western Europe and opened up in Iberia (Figure 6.2). There is significant evidence of population decline and a bottleneck in western Europe (Demars, 1996; Bocquet-Appel & Demars, 2000a; Richards *et al.*, 2000) coinciding with the Solutrean expansion in Iberia. Straus & Winegardner (2000) have estimated site densities for the Atlantic–Cantabrian and Mediterranean regions of Iberia during the Upper Palaeolithic. Giles Pacheco *et al.*'s (2003) data for the corresponding period closely correlated with Straus & Winegardner's (2000) Mediterranean data. Neither was significantly correlated with the Atlantic–Cantabrian data indicating that this latter region has behaved differently in respect to human occupation (Figure 7.8). Such a conclusion is in keeping with Finlayson's (1999) and Finlayson *et al.*'s (2000a) distinction of this region as bioclimatically Euro-Siberian or temperate oceanic, with greater affinities to western Europe that to the rest of Iberia (see next section). Thus, the Aurignacian and Gravettian are much more significant in the Atlantic–Cantabrian region than anywhere to the south. There is also a north–south trend for the Gravettian, which represents a dual effect: (1) the earlier arrival of steppe environments in bioclimatic zones that were closer to those of western Europe; and (2) a distance effect as people took longer to reach southern Iberia. We can contrast the Iberian pattern with that of the more continental Italian and Balkan peninsulas that also happened to be further east and therefore closer to the source of the Aurignacian. In Greece, an Upper Palaeolithic industry with blades with curved back and microliths dated to 40 kyr precedes the Aurignacian (dated at 32 kyr) (Koumouzelis *et al.*, 2001) and suggests local adaptation to changing circumstances in the heterogeneous mid-latitude belt, that we would expect to reach that part of the world sooner than the west, followed by the arrival of the Aurignacians and their slow infiltration of these environments. In Italy, the Aurignacian reaches south to Sicily (Chilardi *et al.*, 1996). These patterns, including the early arrival of the Aurignacian to northern Iberia, contrast with the late or non-arrival of the Aurignacian to southern Iberia.

Another noteworthy difference between northern and southern Iberia is the response to the deglaciation after the LGM. There was a population decline in the Magdalenian followed by a subsequent expansion in the Epipalaeolithic (Figure 7.8). Giles Pacheco *et al.* (2003) interpreted this to mean that the Magdalenian people of southern Iberia were the same as the Solutreans with a primary

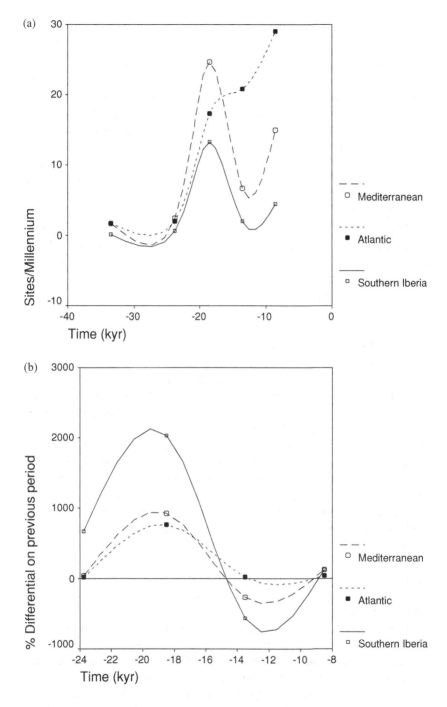

Figure 7.8. Changes in site density during the Upper Palaeolithic in different regions of the Iberian Peninsula. Note the close correspondence between the Mediterranean and the south (Mediterranean bioclimatic areas) and the sharp contrast with the Atlantic (Euro-Siberian bioclimatic areas).

adaptation for steppe conditions. As these environments shrunk so population densities dropped, first in the south. This interpretation is in keeping with the general observation that climate warming is known to affect a northward expanding species' ability to survive in the south of the range (Bennett *et al.*, 1991; Hewitt, 1996, 1999). Response to the climatic warming would have been the use of new technologies, that are evident in the Magdalenian, and an initial tracking of steppe environments. The latter response would take the form of a following of these environments up mountains in the south, and Straus & Winegardner (2000) indeed comment on an increase in mountain sites in the Magdalenian. In the north there was a population expansion and Straus & Winegardner's (2000) results show an increase in site density in the Magdalenian in the Atlantic–Cantabrian region when populations in the south were on the decline. This phenomenon appears to be part of a wider pattern of northward expansion at the end of the LGM (Demars, 1996; Torroni *et al.*, 1998; Bocquet-Appel & Demars, 2000a).

The transition in Iberia

Empirical evidence currently points towards the Iberian Peninsula being one of the last geographical regions of the planet in which Neanderthals survived (Vega-Toscano *et al.*, 1988; Antunes *et al.*, 1989; Zilhao, 1993, 1995, 1996; Raposo & Cardoso, 1998; Finlayson & Giles Pacheco, 2000). More widely, the causes of the extinction of the Neanderthals are unknown although most authors link the disappearance with the arrival of Modern Humans in Europe and western Asia even though evidence of biological superiority of Modern Humans over Neanderthals, to which many authors still subscribe (see 'Comments' in d'Errico *et al.*, 1998), is non-existent as we have seen in this book. At best the logic applied in favour of such superiority rests on the circular reasoning that they (the Moderns) survived and must have therefore been superior (see 'Comments' in d'Errico *et al.*, 1998). Nobody is prepared, it seems, to consider the possibility that the colonisation of Europe by Moderns and the extinction of the Neanderthals may have been independent events, a position that I have advanced in this book.

More specifically in Iberia, attempts have been made to link the extinction of the Neanderthals there with climate change which caused the entry of Modern Humans into the Iberian Neanderthal refuge (Zilhao, 1996; Finlayson & Giles Pacheco, 2000). The Iberian Peninsula is particularly well suited for the study of the 'transition' or 'replacement' (which I prefer to call 'extinction:colonisation processes', a terminology closer to that of existing theory, MacArthur & Wilson, 1967) given a large surface area, biogeographical distinctness and its ecological heterogeneity caused largely by its highly varied relief (Finlayson & Giles Pacheco, 2000; Finlayson *et al.*, 2000a)

The basis of this section is an ecological model which aims to point at possible underlying mechanisms for the extinction of the Neanderthals in the Iberian Peninsula by disentangling the multiplicity of potential variables and analysing the effect of a small number of sufficient parameters (Levins, 1968). By doing this I hope to establish patterns of wider geographical relevance in support of the arguments advanced in this book. Most recently, d'Errico *et al.* (1998) have called for the need to develop models of what happened to the Neanderthals. If indeed we are to resolve this question scientifically (Kuhn, 1970) then we must proceed through the development of testable strategic models and more general tactical ones (May, 1973; Gillman & Hails, 1997) which form the theoretical framework upon which empirical evidence must be evaluated.

For this exercise I have divided the Iberian Peninsula into 273 50 × 50-km Universal Transverse Mercator (UTM) Projection units. A map of the bioclimatic stages of the Iberian Peninsula (Rivas-Martínez, 1987) was superimposed on this grid and a bioclimatic stage was allocated to each unit. In cases where, for reasons of abrupt relief, more than one bioclimatic stage occurred in a square the stage which was judged to cover the greatest surface area of the square was allocated to the square. Four bioclimatic stages were identified, but the oro- and crioro-Mediterranean stages (Rivas-Martínez, 1981, 1987) are too reduced in area and restricted to certain mountain peaks to be significant at the resolution of the model (Figure 7.9a). The four stages I have used (following Rivas-Martínez, 1981, 1987) are: (1) thermo-Mediterranean, characterised by mean annual temperatures (T) between 17 and 19 °C, mean minima of the coldest month (m) between 4 and 10 °C and mean maxima (M) of the coldest month between 14 and 18 °C; (2) meso-Mediterranean, characterised by T between 13 and 17 °C, m between −1 and 4 °C and M between 9 and 14 °C; (3) supra-Mediterranean, characterised by T between 8 and 13 °C, m between −4 and −1 °C and M between 2 and 9 °C; and (4) Euro-Siberian characterised by T between <3 and 10 °C, m between <−8 and 0 °C and M between <0 and >8 °C. Elements of the oro- and crioro-Mediterranean stages would have fallen within supra-Mediterranean stages. The characteristics of these stages are: (1) oro-Mediterranean, T between 4 and 8 °C, m between −7 and −4 °C and M between 0 and 2 °C; and (2) crioro-Mediterranean, $T < 4$ °C, m $< −7$ °C and $M < 0$ °C.

Using the range in T of the different bioclimatic stages, I have calculated how the proportion of 50 × 50-km units allocated to each bioclimatic stage would vary with a progressive drop in T from present-day to 10 °C below present at intervals of 1 °C. The results (Figure 7.9b) indicate that Mediterranean bioclimatic stages would not disappear altogether at $T - 10$ °C although only the high

(a)

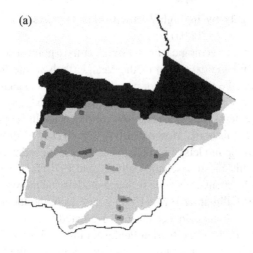

(b)

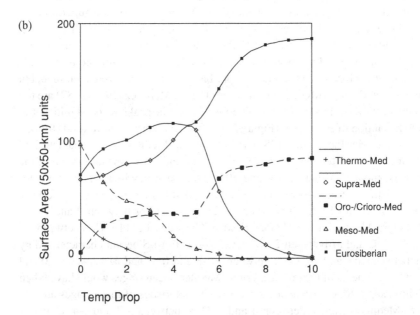

Figure 7.9. Predicted changes in the distribution of Iberian bioclimatic stages with decreasing annual mean temperatures (T). (a) Present day distribution. (b) Curves showing changes in bioclimatic stages with decreasing T at 1° intervals.
(c) Distribution of bioclimatic stages at T −3 °C. (d) Distribution of bioclimatic stages at T −9 °C. Black: Euro-Siberian; dark grey: oro/crioro-Mediterranean; medium grey: supra-Mediterranean; pale grey: meso-Mediterranean; white: thermo-Mediterranean.

(c)

(d)

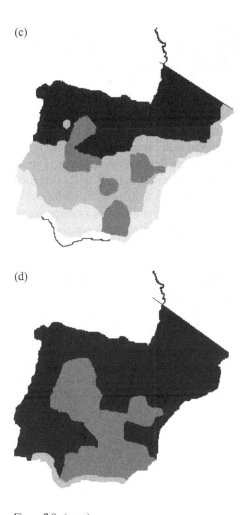

Figure 7.9. (*cont.*)

montane components (oro- and crioro-) would remain. The thermo-Mediterr-anean Stage disappears at $T -3\,°C$; the meso-Mediterranean at $T -7\,°C$; and the supra-Mediterranean at $T -10\,°C$ although the area covered by this stage actually increased between $T -1\,°C$ and $T -5\,°C$. The oro- and crioro-Mediterr-anean and the Euro-Siberian stages progressively expanded in area with declin-ing T. The range of T examined encompasses the conditions that would have been met in Iberia in OIS 3 and 2. In the Massif Central in France, warm OIS 3 events were $T -4\,°C$ and cold events down to between $T -9\,°C$ and $T-11\,°C$. (Guiot *et al.*, 1989). Using the results from Figure 7.9(b), I have illustrated in Figure 7.9(c) the expected distribution of the Iberian bioclimatic stages at $T -3\,°C$, a situation that may have been frequent in OIS 3. It reflects the elimination of the thermo-Mediterranean stage and the growth of the montane Mediterranean stages. In Figure 7.9(d) I represent an extreme situation at T $-9\,°C$, resembling the colder late stages of OIS 3 and the conditions at the LGM. The dominance of Mediterranean montane stages and the southward expansion of the Euro-Siberian stages are evident. This pattern fits well with the observations of environmental change in Iberia that I described in Chapter 6.

In the density-dependent population dynamic models that I introduce here, growth and range expansion/contraction are related to bioclimatic and geographical factors. It is important to stress that I use the terms warm-adapted, cold-adapted, etc. to convey principally the direction of response to climate-induced habitat change. Thus, a population that reacts positively to an increase in preferred habitat and that habitat increases as conditions become colder, is said to be cold-adapted, and so on. In each case the starting point of the model was selected at 58 kyr for Neanderthal populations and 40 kyr for Moderns. The first was chosen to cover the spectrum of climatic change of OIS 3 (e.g. van Andel, 1998; van Andel & Tzedakis, 1998); the second was considered to be a close approximation to the time when Neanderthal and Modern populations were sympatric in the north of the Iberian Peninsula (Mellars, 1996) and close to the point of first appearance of Moderns in Iberia. From these points population levels could be estimated forward or backward in time. The assigned ages are therefore model-dependent throughout. Depending on whether the simulation being run assumed that the population showed a preference for warm or cold conditions the starting population size was varied for each bioclimatic stage. Thus, for Neanderthal simulations at 58 kyr the starting population sizes allocated for each of the 273 cells were set at 100, 75, 50 and 25 population units ranked by bioclimatic stages. Thus, if the model considered the population to be warm-adapted the thermo-Mediterranean stages would have a starting population level of 100 and the Euro-Siberian a starting population level of 25 and vice-versa. Reversing the order of the initial population had a negligible effect on the model outcome. Figure 7.10 shows, for example, that the extinction of the warm-adapted Neanderthal population occurred within a 1-kyr period with the two extreme starting population sizes. Sensitivity analysis further emphasised the model's robusticity, the starting population contributing 1.4% to the variance of the outcome. The effect of stochasticity on the model outcomes was tested by running Monte Carlo simulations (1000 trials). For the warm-adapted Neanderthal model the population became extinct at the simulated outcome time in the deterministic model on 47.3% of the trials. For the remainder, the population size had fallen to under 3 and was thus in imminent danger of

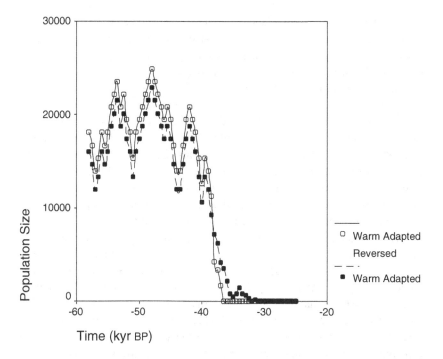

Figure 7.10. Simulated evolution of the Iberian Neanderthal population with starting populations described in the text (warm adapted) and with starting populations reversed (warm adapted reversed).

extinction. For the cold-adapted Moderns model the population size at 25-kyr varied between 29 780 and 56 744 and never became extinct. The model assumed that the bioclimatic differences between the four stages were maintained relative to each other as climatic conditions fluctuated. The criteria for Modern Human populations differed. These always started at a level of 50 in the Euro-Siberian zone where they originated and where it was assumed they had not reached the environmental carrying capacity. In no simulation were Modern Human populations established in the other bioclimatic stages at 40 kyr.

The time interval chosen for the simulations was 0.5 kyr. Temperatures were estimated into four relative categories for each interval from the ^{18}O Greenland Ice Cores (GRIP, 1993): warm (10.6% of the intervals); moderately warm (22.7%); moderately cold (24.2%); and cold (42.4%). Thus, cold events dominated OIS 3 (66.6%; $n = 67$). The proportion of cold intervals increased towards the end of OIS 3: 58–50-kyr, 47.06% cold, $n = 17$; 50–40-kyr, 60% cold, $n = 20$; 40–30-kyr, 80% cold, $n = 20$; 30–25-kyr, 80% cold, $n = 10$. If a population was considered to be warm-adapted then it was allocated a growth rate per cell of 10 population units for each warm period and of 5 for each moderately warm period, a decline rate of 5 for each moderately cold period and of 10 for each cold period. The reverse applied to a cold-adapted population. A warm-tolerant population would maintain the same population level during warm periods but would decline as above in cold events. The reverse applied to a cold-tolerant population. Growth rates were the same across all bioclimatic stages. A growth rate of 10 units per time interval would, if the units were equated to people (see below) mean a growth by 2 per 100 years, a measure that is considered conservative for an expanding population. A population reaching a population level in a cell below 0 population units was considered extinct. Recovery into that cell would only occur where an adjacent cell still retained a population of the same species and only if climatic conditions were favourable for growth. Modern populations could only spread into adjacent cells under favourable conditions. Movement from one cell to another only occurred from populated cells to adjacent cells during favourable climatic conditions and between consecutive bioclimatic stages and at 0.5-kyr intervals. In cases where adjacent cells were more than one bioclimatic stage apart, colonisation was only permitted after two consecutive favourable time intervals. Thus, for example, if a populated cell belonged to the supra-Mediterranean stage, then an adjacent meso-Mediterranean cell could be colonised at the next time interval by an advancing cold-adapted Modern population provided that conditions in that interval were cold. If the adjacent cell was thermo-Mediterranean instead, then the colonisation would require two successive cold intervals.

Four scenarios were tested for Neanderthals and two for Moderns. These were considered to provide contrasting situations and clearly other models could be generated to test more extreme situations. It is considered that the scenarios tested serve to highlight the main testable predictions of the models. The four Neanderthal scenarios tested the populations to be cold-adapted, cold-tolerant, warm-tolerant and warm-adapted (Figure 7.11). For Moderns cold-adapted and warm-adapted scenarios were tested. These models were intra-specific density-dependent models. In other words the populations of the two forms fluctuated in response to environmental conditions and their own starting population levels but were unaffected by the presence of the other form. Population sizes per cell were capped at twice the maximum number of starting units (i.e. 200) irrespective of bioclimatic stage to reflect a measure of environmental carrying capacity. If we were to equate population units to people in the landscape, then 100 units in a 50×50-km square would represent a density of $0.04/\text{km}^2$ or 4 per 100 km^2; 200 units would represent $0.08/\text{km}^2$. In Africa, Hadza population density in an area of 2500 km^2 (therefore equivalent to one cell in the Iberian models) was between 600 and 800, that is 0.24-$0.32/\text{km}^2$ (O'Connell &

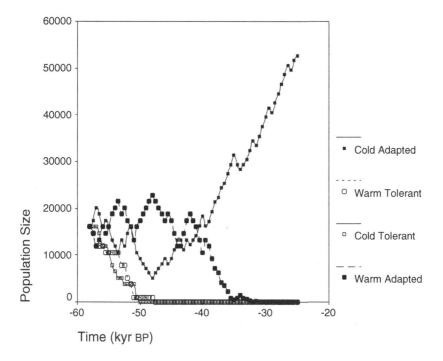

Figure 7.11. Contrasted fates of the simulated Neanderthal population under four different scenarios. See text for details.

Hawkes, 1988), and !Kung San density in Dobe in the Kalahari was 466 in 9000 km^2, that is 0.05/km^2 (Lee, 1979). Given that the density of most contemporary hunter–gatherer populations ranges between 0.01 and 0.4/km^2 (see, for example, Diamond, 1991) these model estimates are conservative and within expected limits. In order to test the effect of interspecific competition a scenario was generated in which the presence of Modern Humans in the same cell as Neanderthals further depressed the Neanderthal population by an additional 5 population units per time period of contact. This is a significant competitive coefficient being equivalent to between 0.5 and 1.0 of the population growth rate in favourable conditions and exceeding the growth rate in unfavourable conditions.

Finally, in order to test longer time-scale implications, models spanning the time period 130–0 kyr were generated for Neanderthal populations. In these cases the time interval was set at 2 kyr but the rate of population change was kept as in the earlier models. The climate estimates were based on Imbrie *et al.* (1992).

The underlying conclusion of all the models was invariant of the degree of climate-induced population fluctuation modelled or of the starting population size: given the nature of the climatic oscillations of OIS 3, warm-adapted populations always tended towards cumulative decline, and cold-adapted ones always grew in size and never became extinct, supporting Finlayson *et al.*'s (2000a) global model of human colonisation and Neanderthal extinction.

Of the four climatic models generated for Iberian Neanderthal populations the warm-adapted, cold-intolerant model best fitted the empirical evidence (Figure 7.11). The cold-adapted, warm-intolerant model generated an expanding Neanderthal population in Iberia with time and the population was still growing at the cut-off point of 25 kyr. The warm-tolerant and cold-tolerant models produced population extinctions at 44 and 42 kyr respectively, far too rapid to fit the empirical data. Figure 7.11 illustrates the evolution of the warm-adapted, cold-intolerant, Iberian Neanderthal population from 58 kyr to its extinction at 31 kyr. For Moderns, the warm-adapted, cold-intolerant, model generated the extinction of the Iberian population at 36 kyr and was thus unrealistic. The cold-adapted, warm-intolerant model for Moderns on the other hand produced an expanding population which eventually colonised the entire Iberian Peninsula, a pattern consistent with more generalised models of human population growth (Ambrose, 1998). The evolution of the warm-adapted Neanderthal and cold-adapted Modern Human Iberian populations from 40 kyr to 25 kyr is illustrated in Figure 7.12. According to this model there is a protracted period of 9 kyr (40–31 kyr) during which both populations occur within the Iberian Peninsula.

Five cells were selected for analysis of the evolution of local populations within Iberia and these were chosen to correspond to bioclimatically distinct units within which empirical archaeological evidence existed. I shall call these cells by the archaeological site which they represent: Gibraltar, Carihuela, Caldeirao, Ermitons and El Castillo (Figure 7.13). The five sites show distinctive patterns of population evolution (Figure 7.13). The Gibraltar population persists longest and actually grows on two occasions during the period, at 34–35 kyr and 31.5 kyr, the latter just before its final extinction at 31 kyr (Figure 7.13a). At 32 kyr the population is briefly regionally extinct and there are no human populations in Gibraltar. Moderns do not arrive until after the Neanderthals have become extinct but the population grows rapidly from 30 kyr (Figure 7.13a). This model therefore predicts no overlap of the two populations in Gibraltar. The situation in the mountains to the north-east, at Carihuela, is similar (Figure 7.13b). Here the Neanderthal population does not exhibit the temporary recovery of the Gibraltar population and the population becomes finally extinct at 33.5 kyr, 2.5 kyr before the Gibraltar population. There is also an earlier period of regional extinction, between 35.5 and 34.5 kyr, during which time there are no

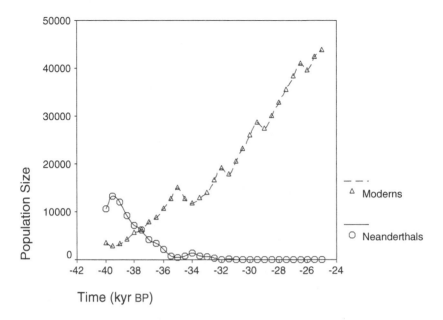

Figure 7.12. Simulated evolution of the Neanderthal and Modern Iberian populations.

human populations in Carihuela (Figure 7.13b). Moderns arrive at 32 kyr, 2 kyr before they reach Gibraltar, but there is no overlap with the Neanderthals (Figure 7.13b). The pattern in Caldeirao (Portugal) to the north-west is again different. The Neanderthal population has a similar recovery capacity to the Gibraltar population with increases at 35–34 kyr and 31.5 kyr, when Moderns are already in the area (Figure 7.13c). The Neanderthal population is extinct at 32.5 kyr but there is a re-entry into the area at 31.5 kyr with the final extinction at 31 kyr, at the same time as at Gibraltar that is further south and due to the bioclimatic situation of this site (Figure 7.9). The Moderns, however, arrive much earlier (by 35.5 kyr) so there is a period of 3 kyr during which the two populations overlap regionally, the highest of all (Figure 7.13c). It is noteworthy that Bocquet-Appel & Demars (2000b), using a different modelling procedure, reached a similar conclusion. In the north, the site at Ermitons is within the Mediterranean coastal region and the Neanderthal population survives until 36 kyr (Figure 7.13d). The model predicts a re-entry of Neanderthal elements into the area briefly at 34 kyr, by which time the Moderns are well established, and the final extinction takes place at 33.5 kyr. The Moderns arrive early (at 38.5 kyr) so we observe here a long period of regional overlap of the two populations, predicted at 2.5 kyr (Figure 7.13d). Finally, the pattern at El Castillo, within

(a)

(b)

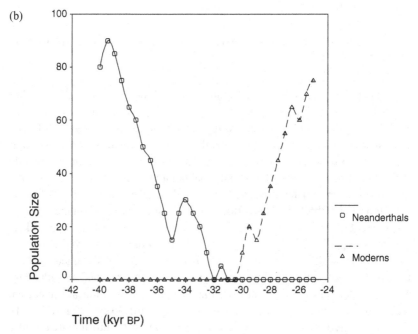

Figure 7.13. (a) Regions selected for analysis based on known sites; (b) simulated evolution of Neanderthal and Modern populations in Gibraltar; (c) Carihuela; (d) Caldeirao; (e) Ermitons; and (f) El Castillo.

(c)

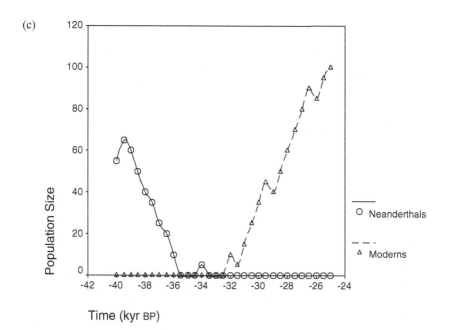

(d)

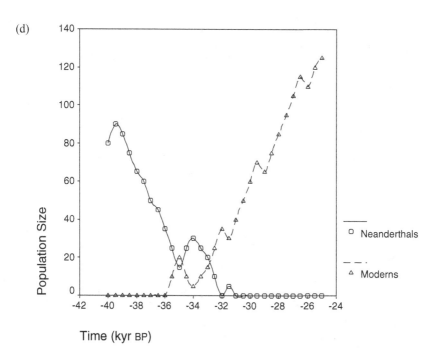

Figure 7.13. (*cont.*)

(e)

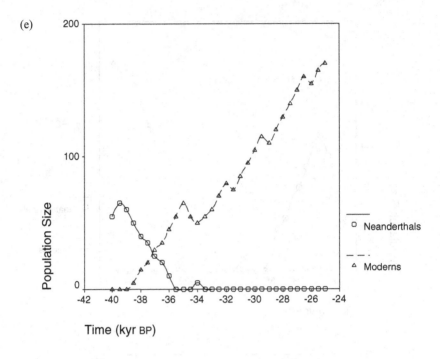

(f)

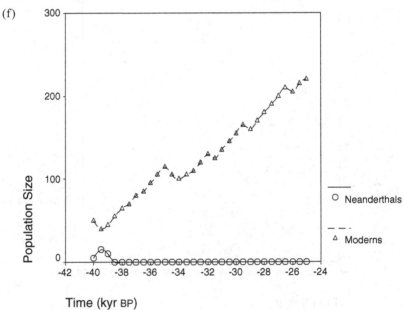

Figure 7.13. (*cont.*)

the Euro-Siberian bioclimatic stage, is very distinct (Figure 7.13e). The Neanderthals are extinct by 38.5 kyr and Moderns are present from 40 kyr, always at higher population levels than the Neanderthals. The model predicts a regional overlap of 1.5 kyr.

Figure 7.14 illustrates the geographical distribution patterns of Neanderthal and Modern Human populations in Iberia at 39 kyr and 33 kyr as generated

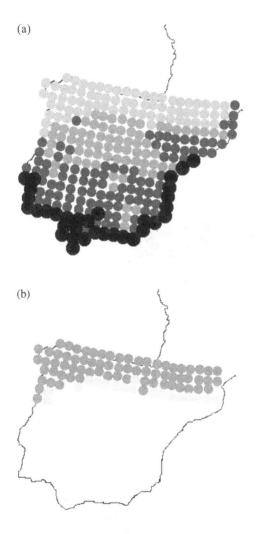

Figure 7.14. Simulated geographical distribution of Neanderthal and Modern populations. (a) Neanderthals at 39 kyr; (b) Moderns at 39 kyr; (c) Neanderthals at 33 kyr; and (d) Moderns at 33 kyr. Darkness and circle size reflect modelled densities.

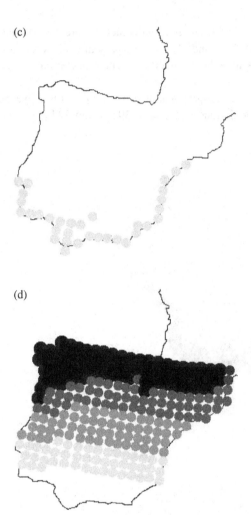

Figure 7.14. (*cont.*)

by the model. The pattern shows a well-established Neanderthal population in Iberia at 39 kyr with the main areas of population density being within the thermo-Mediterranean stages (Figure 7.14a). The Moderns are restricted to the north and appear thinly spread on the ground (Figure 7.14b). At 33 kyr, only 6 kyr later, the Neanderthals are thinly spread and restricted to the thermo-Mediterranean stages (Figure 7.14c). The Moderns are firmly established in the north and Moderns have spread to almost the whole of the peninsula except the southernmost regions (Figure 7.14d).

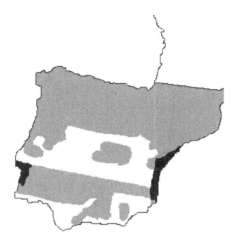

Figure 7.15. Predicted periods of overlap between Neanderthals and Moderns. Black, 2.5 − 4 kyr; grey, 0 − 2.5 kyr; white, no overlap.

The predicted lengths of regional overlap are illustrated in Figure 7.15. The mosaic nature of the distribution results from the heterogeneous bioclimatic landscapes of Iberia and the time lags associated with the progressive southward spread of Moderns. Overlaps ranged from no overlap to maxima of 4 kyr at the regional spatial scale even though the two forms were present together in Iberia, as we have seen, for 9 kyr. There are large areas of the central Meseta and of southern Iberia where no overlap is expected and in which the extinction of the Neanderthal populations cannot be related to the presence of Moderns. I have commented elsewhere (Finlayson, 1999) that the probability of hybridisation between Neanderthals and Moderns ought to be greatest in areas of high temporal overlap. The conclusion is that the effect of competition at the simulated level is minor and only changed the outcome of the Neanderthal population by a small degree. The model suggests that a huge advantage of the Modern population over the Neanderthal would be required for competition to have an overriding effect over the climate. Clearly, there is no competitive effect in certain areas (e.g. Gibraltar and Carihuela) where the two populations do not meet at all.

A possible criticism of the decline of Neanderthal populations in response to cold, as we have seen in this chapter, is that the same populations would have had to undergo even colder episodes earlier in their history which they survived. In order to answer this question I extrapolated the Neanderthal warm-adapted, cold-intolerant model back in time from the established population sizes of the intial model while maintaining the same criteria for rate of population change in

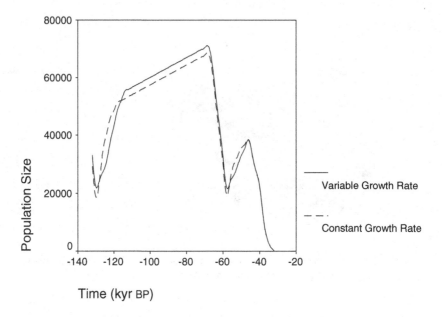

Figure 7.16. Simulated evolution of the Iberian Neanderthal population since the last interglacial.

response to climatic factors. I then upgraded the model further by introducing variations in recovery rate during warm intervals depending on the bioclimatic stage (i.e. from increments of 5 in the Euro-Siberian stage to 20 in the thermo-Mediterranean). The model thus supplants a uniform growth rate by a variable growth rate dependent on the bioclimatic stage. The difference between the 'standard' model and the upgrade was one of degree but both produced a very similar result (Figure 7.16). The model shows that the Iberian population would have reached its heyday at the end of the last interglacial (OIS 5a) which would have been when ecotonal landscapes would have peaked. At its height the Iberian modelled population was of the order of 70000. This would translate to a population density of $0.1/km^2$, well within the hunter–gatherer range described earlier. With some intervals of minor recovery, from then on the population was in a steady and consistent decline towards extinction (Figure 7.16). The decline was particularly abrupt during the severe conditions of OIS 4 (Mellars, 1996) but the population was able to recover from this crash, and consequent bottleneck, although the Stage 5a population level could not be attained again. The relatively mild and variable climate of early OIS 3 permitted a partial recovery. The cooling towards the end of this stage, though not as severe as earlier events, was enough to seal the fate of the population since the population level at the start of the crash was already depressed from the effects of OIS 4. It is the cumulative effect of a number of cold events that appears to account for the extinction of the Neanderthals. A comparison of the pattern for the southern site of Gibraltar and the northern site of El Castillo illustrated how much closer the northern population came to extinction at the end of OIS 4 than the southern population. By extrapolation it can be inferred that more northerly European populations would have been severely affected by cooling events after the last interglacial and many probably became extinct. Subsequent spreads into western Europe would have involved populations from the Iberian refugium. The south-western region of France is interesting in this respect. Some Neanderthal populations may have been able to survive here throughout as they did in northern Spain. The proximity of this region to the Iberian refugium would have also meant rapid recolonisation during warm events. Its more northerly position and proximity to the European Plain would have also made it more vulnerable to climatic instability and would have also permitted an earlier colonisation by Moderns than further south. The relationship between Chatelperronian technology, associated with interstadial conditions around 36 kyr, followed by Aurignacian with the onset of glacial conditions at 34 kyr (Mellars, 1996) is in accord with these predictions.

Of the series of models that were run the one that I have presented in detail appears to be the one which best fits the existing empirical data. The basis of any good model is that it should make testable predictions which should allow progress to be made through falsification or confirmation. The following testable predictions are derived from the model:

(1) The geographical pattern of Neanderthal extinction in Iberia is not a straight north to south process but is related to bioclimatic stages. Specific predictions can be made for each 50×50-km UTM unit.

(2) The spread of Moderns from the north to the southernmost points of the Iberian Peninsula takes 10 kyr.

(3) It is unnecessary to postulate a static barrier (e.g. an Ebro Frontier; Zilhao, 1996) which divides Neanderthal and Modern Human populations for a protracted period. Delays between arrival of Moderns in different parts of the Iberian Peninsula are predicted by the model.

(4) Neanderthals and Moderns never met in some parts of the Iberian Peninsula. The pattern of overlaps is not north–south but is in the form of a mosaic. Geographical areas are defined by degree of overlap.

(5) The effect of stochastic extinction and re-colonisation of local populations within regions, independently of the situation of the global population, can create apparently opposing patterns in specific localities.

(6) The Neanderthal population of Iberia, and by extrapolation that of western Europe, reached its maximum level during OIS 5a after which numbers progressively dwindled towards eventual extinction.

(7)   If Upper Palaeolithic technologies evolved independently in Neanderthals
and were not the product of acculturation (d'Errico *et al.*, 1998) then such
technologies should be found in Neanderthal regions of Iberia where there
was no contact with Modern Humans.

In this section I have shown how a local population effect may run counter
to a regional or more global population trend as part of the dynamics of the
system. Thus, within a geographical region opposing population trends may
result in different local populations and stochastic effects can be particularly
important in small populations. These observations raise the question of the
level of confidence which may be attached to the interpretation of patterns of
population spread, of transition or replacement of populations, from the study
of a few discrete and geographically widely-separated sites.

With this caveat in mind I now attempt to correlate the regional predictions
of the model for the five selected areas with the existing evidence from the five
representative sites. Since the dates used in the simulation are model-dependent
we can only compare site patterns relative to each other. For the end of the Middle Palaeolithic the model predicts the following sequence of disappearance:
Castillo, Ermitons/Carihuela; and Caldeirao/Gibraltar. The data in the literature
indicate the sequence Castillo, Ermitons, Gibraltar, Caldeirao/Carihuela (Vega-
Toscano *et al.*, 1988; Vega-Toscano, 1990; Cabrera & de Quiros, 1996; Maroto
et al., 1996; d'Errico *et al.*, 1998; Stringer *et al.*, 2000). For the start of the
Upper Palaeolithic the model predicts the following sequence: Castillo, Ermitons, Caldeirao, Carihuela, Gibraltar. The data in the literature indicate Castillo,
Ermitons (in fact sites in the same area), Caldeirao/Gibraltar/Carihuela (Vega-
Toscano *et al.*, 1988; Vega-Toscano, 1990; Cabrera Valdes & de Quiros, 1996;
Maroto *et al.*, 1996; d'Errico *et al.*, 1998; Stringer *et al.*, 2000). Given the
difficulties described above and the different dating techniques applied at the
different sites the observed trends would seem to match the model predictions
very closely.

The coastal region of Cataluña would, according to the model, be a zone
of extended overlap. Maroto *et al.* (1996) record that 'As far as Catalonia is
concerned, the passage from Middle to Upper Palaeolithic is characterised
by two phenomena which, though they may at first seem contradictory, are
quite compatible: abrupt replacement and coexistence.' This is exactly what
the model predicts. In the Portuguese sites (Caldeirao and others) Raposo &
Cardoso's (1998) data indicate some sites in which the Upper Palaeolithic and
Mousterian levels are in close approximation which could indicate proximity in
time and the Lagar Velho site (Duarte *et al.*, 1999), as we have seen, is strongly
indicative. At Gorham's Cave, Gibraltar, there is nothing so far that indicates
any temporal continuity between Middle and Upper Palaeolithic. This limited

evidence is suggestive but more work will be needed to test the prediction of a variable mosaic of overlap zones, including regions where there was no contact at all, within Iberia

Given an arrival into the open areas of Europe the Moderns would have found little ecological or physical resistance to rapid spread and such progress would only have been checked when reaching topographically or ecologically distinct zones (in addition they probably did not meet many Neanderthals along the way). Such areas would have included Alpine regions and other high mountain blocks. The Iberian Peninsula, with its distinct biogeographical character, would have provided such a check until climatic conditions permitted its rapid penetration as the ecological barriers became increasingly diffuse. It is unlikely that the spread of the Moderns would have come to a complete halt and the idea of a barrier in the region of the Ebro (Zilhao, 1996; d'Errico *et al.*, 1998) may only have been an illusion given the discontinuous distribution of archaeological sites to its south and the difficulty of detecting initially small colonising populations from such sites as exist. In fact, in areas of long-term regional sympatry patterns such as those described for Iberian zones of overlap would be expected with at least partly stochastically-driven local extinctions, replacements and even apparent trend reversals occurring. Such variability has been described for Middle Eastern sites (Stringer & Gamble, 1993; Bar-Yosef, 1994) and may reflect a combination of the expansion of the southern limit of the Neanderthal range in response to climatic deterioration and stochastic processes expected in small metapopulations (Harrison, 1991; Foley, 1997; Hanski & Gilpin, 1997). Stochastic effects would have also affected the Modern population which was at low densities in North Africa and the Middle East during the early last glacial period (Ambrose, 1998). In general, late Neanderthal populations would be expected from heterogeneous environmental situations, at the regional scale, where climatic characteristics maintained refuges of 'non-steppe/tundra' vegetation. Such areas, which at low latitudes would be expected to have the highest mammal species richness (Kerr & Packer, 1997) and the greatest tree diversity (McGlone, 1996), would have included, as we have seen, the three Mediterranean peninsulas – Iberia (Vega-Toscano *et al.*, 1988; Antunes *et al.*, 1989; Zilhao, 1993, 1995, 1996; d'Errico *et al.*, 1998; Raposo & Cardoso, 1998; Finlayson & Giles Pacheco, 2000), Italy and the Balkans (Karavanic & Smith, 1998) – the Middle East (Bar-Yosef, 1998), areas on the edge of the Russian Plain (Soffer, 1989) and, to a lesser degree, the heterogeneous landscapes of south-western France (Mellars, 1996). In this respect the more northerly distribution of steppe and tundra in eastern Europe and western Asia (Chapter 6) than in western Europe (Mellars, 1996) could account for the late survival of Neanderthals around the Black Sea. Such heterogeneous landscapes would have reached their maximal extent at the start and end of interglacial conditions

(Mellars, 1996) and probably also locally during full interglacial conditions, when forests spread across much of Europe, through the clearing action of megaherbivores (Owen-Smith, 1988).

In Gibraltar, the emerging pattern is that of a Neanderthal population well tuned into the seasonal resource cycles within their home ranges (Finlayson & Giles Pacheco, 2000; Finlayson *et al.*, 2000a). Neanderthals were omnivorous. Opportunistic scavenging (Stiner & Kuhn, 1992; Mareau, 1998) would be an additional tactic within such a system. Such a strategy would only be successful in environmentally heterogeneous regions and would have reduced risks which could have arisen from over-dependence on a single or small number of prey types (Gamble, 1995). A similar situation applied in Italy (Stiner, 1994; Kuhn, 1995), south-western France (Mellars, 1996) and the Middle East (Stringer & Gamble, 1993). The tactics of survival in regionally homogeneous landscapes (such as the grass savannahs or the steppes) would have been very different, requiring a different kind of social organisation, planning and long-range movements (Chapter 5). A population with such a social structure (Bar-Yosef, 2000) would have been predisposed to success. If we add to this the short duration of the northern summers and the short daylengths of the northern winters we see that a very different strategy would have been required to survive in the environments of the Eurasian Plain during glacials. Such a strategy must have included an improved monitoring of distant environments (Bar-Yosef, 2000). The Moderns did eventually spread into the kinds of heterogeneous landscapes occupied by the Neanderthals but they may have done so largely once the Neanderthals had left these or after the optimal environments of the Moderns had been filled. The occupation of mountain environments by Moderns occurred relatively late (Gamble, 1993).

Thus Neanderthals would appear to behave in the traditional sense as K-selected whereas the expanding Moderns (with a growing population and largely not at carrying capacity) would fit better an r-selected model (Chapter 5; MacArthur & Wilson, 1967; MacArthur, 1984). McGlone (1996) has suggested three ways in which temperate and boreal trees ran the risk of being eliminated from large portions of their range during a glacial–interglacial cycle: (1) through not colonising sufficient sites during their period of peak abundance; (2) through elimination by stochastic effects or competition in long periods of isolation; and (3) through failure to track suitable climates through migration during periods of rapid change. The analogy for a K-strategy population of Neanderthals is self-evident.

The main implication of all this then is that the arguments advanced in this book that Neanderthal extinction and Modern colonisation can be understood without recourse to interpretations that require the biological superiority of one form over the other is reinforced. Quite simply, the Modern phenotype was better suited to the variable and cooling conditions of OIS 3 and OIS 2

Europe than the Neanderthal phenotype was through responses to the expansion of the biotopes with vegetation structural characteristics akin to those in which that phenotype is likely to have evolved (Foley, 1987). The Neanderthals, according to the model presented here, had been declining in Europe since their heyday in OIS 5a, and especially during OIS 4, always in the absence of Moderns.

The model establishes one point conclusively, whether or not Moderns and Neanderthals interacted and whether such interactions were competitively unequal, such interactions are not a prerequisite for Neanderthal extinction. It is perfectly plausible to explain the Neanderthal extinction as a result of environmental change caused by climate as we have seen. As such there would be areas of Iberia which would have been colonised by Moderns after Neanderthals had disappeared. It is also possible to observe scenarios in which a declining Neanderthal population and a growing Modern population could coexist and inter-breed within the same geographical region for up to 4 kyr. These areas would have been of reduced extent (1.5% of the Iberian surface area in this model) so it is therefore unlikely that large-scale hybridisation would have been the rule (Finlayson, 1999). A real measure of the degree of interaction between the two forms would result from the study of independent cultural evolution vs acculturation in regions where the two forms overlapped and others where they did not at all.

Synthesis

I now summarise the events leading to the Modern colonisation of Eurasia and the Neanderthal extinction.

Human adaptation

Humans during the course of the Pleistocene adapted to those features that were relatively stable. These were the availability of medium-sized mammalian herbivores and the topography of the terrain. Neanderthals adapted to hunting mammalian herbivores in broken terrain. Moderns adapted to hunting mammalian herbivores over flat terrain.

Instability

Increasing climatic instability fragmented the tropical African rainforests and African hominids developed a biology consistent with the exploitation of patchy environments. Behavioural plasticity and dispersal ability were favoured. The Moderns were the ultimate expression. The Neanderthals were also behaviourally plastic, having descended from the common ancestor with the Moderns. They coped with instability by extending the range of resources consumed over a small area. Dispersal ability was reduced in comparison with the Moderns.

Habitat tracking

Habitats moved with climate change. Moderns tracked intermediate and open habitats and expanded from Africa. The expansion of open habitats in Eurasia favoured this population. Neanderthals tracked intermediate habitats. As the rate of change accelerated Neanderthals attempted to adapt behaviourally but their morphology was a constraint in open landscapes. As their optimal habitats contracted their range was fragmented. With low dispersal ability they became extinct.

The Modern Human super-organism

Moderns resolved instability through the unified action of teams. Social behaviour, language and symbolism bound teams together and enhanced information flow. Modern Human groups coverted unpredictability into predictability by a variety of behavioural means. Behavioural adaptation, especially through culture and technology, permitted them to ride the increasingly frequent environmental oscillations. Moderns were also stressed during the LGM but they managed to hold out and recolonised Eurasia and the world.

8 *The survival of the weakest*

The thrust of this book is that pressures and stresses on peripheral populations in areas of population growth have driven the changes that have marked human biological and cultural evolution. It has been the stressed populations that have been the most innovative (e.g. Fitzhugh, 2001). Hominids have responded to the increasing instability with risk reduction responses that fit under the umbrella of increasing the spatio-temporal scale of operation. Correlates of increase in scale have included, as we saw in Chapter 5, an increase in dispersal ability, home range size, group size, neocortex size, the complexity of social behaviour, symbolism, efficient and mobile tool kits, and gracile morphology. Since global climate and environments have become increasingly unstable over the last two million years those marginal populations that adapted to local stresses in these ways were able to turn disadvantage into advantage each time conditions deteriorated or became less stable. These adaptations would have evolved in peripheral populations that perceived marginal landscapes as spatially heterogeneous and therefore spatially risky. These adaptations to exploiting patchy landscapes then became advantageous in situations of increasing temporal heterogeneity, that is in situations that were perceived as temporally risky. We can therefore understand these adaptations as evolving through a normal process of natural selection and we do not need to invoke alternative mechanisms, such as variability selection (Potts, 1996a, b, 1998), to explain the observed patterns and trends. Most of these adaptations would have been behavioural. However, directional trends towards increasing patchiness also permitted morphological adaptations such as limb morphology and neocortex size. Temporary, low-scale, climatic and environmental improvements would have reversed the situation. We do not know how many innovations subsequently became 'extinct' as no trace of them would have been left in the archaeological record. The apparent halt of the progress of the early Moderns in the Middle East after 100 kyr may be an example of such a reversal. Had the global situation changed towards increasing stability the picture today may well have been very different as conservative behaviour would have been favoured over innovation. Let us briefly look at large scale technological change to illustrate this point.

Technological innovation

Technology is the result of: (a) ecology, in response to the environment; and (b) history, the product of traditions and their diffusion (Chapter 5). The changes that we observe in the archaeological record, that take us from Mode 1 to Mode 4 technology, do not reflect a linear evolution. Reversals, in cases of the discarding of more sophisticated technology in favour of more rudimentary forms are not infrequent. The adoption of a particular technology is the balance between its costs and the benefits gained from its use. A costly technology may be adopted if it brings gains that far exceed the costs. If circumstances change and the benefits are reduced, the particular technology may be discarded in favour of a less costly form. The apparent increase in technological complexity through time is a reflection of cumulative historical effects and not of a linear, progressive, technological evolution. What we do expect to observe through time, in a world of increasing instability, is the development of technologies that were increasingly costly to make but that were also increasingly beneficial in unstable circumstances. What we would expect to increase is their efficiency – that is, the differential between the energetic costs of manufacture and the benefits derived from their use. I propose that the trend from Mode 1 to Mode 4 is one that goes from low cost/low benefit/low efficiency technologies towards high cost/high benefit/high efficiency ones. We see this degree of sophistication in the late Upper Palaeolithic technology, such as the light and highly specialised microblades of north-east Asia and north-west North America that facilitated the rapid re-colonisation of Siberia after the Last Glacial Maximum (LGM) around 18 kyr (Goebel *et al.*, 2000).

Mode 1, the earliest technology that was employed by the early forms of *Homo* (attributed to *ergaster, erectus, antecessor*), was an economical and local technology. There was no requirement to cover large distances to obtain particular raw materials. Tools were made *in situ* and discarded once used. In the particular circumstances of early Pleistocene Africa, a relatively stable climate but more unstable than previously, this tool kit was advantageous in permitting greater options of resource exploitation. The hominids that practised this technology thrived and Mode 1 spread, permitting the bearers to breach the tropical barrier for the first time. The process of spread was slow. By the time Mode 1 reached Eurasia it was in an evolved form (Carbonell *et al.*, 1999a).

The next technological mode to appear was Mode 2, associated with the production of bifacial tools and with early (*ergaster*) and evolved (*heidelbergensis*) forms. Mode 2 had no clear advantage over Mode 1 as the two survived in parallel for 1 Myr. We have no knowledge of whether groups of hominids alternated between these two technological modes depending on ecological circumstances. Because Mode 2 required particular types of raw materials, which required specific searches that had not been needed for Mode 1 and because there were costs involved in manufacture, we have to assume that this was a more costly technology than Mode 1. In the type of conditions in which Mode 1 evolved, the more costly Mode 2 would have been at a disadvantage which is probably why it took some time for it to become widespread and why Mode 1 was often used instead when the two options were available. How would Mode 2 have originated? In my view Mode 2 would have arisen in peripheral tropical African populations that were more stressed than core populations using Mode 1. As Mode 1 populations expanded it would have been the peripheral populations, in sub-optimal habitats, that would have had to increase their scale of operation to reduce risk of extinction. Living in increasingly patchy landscapes in terms of their preferred resources these populations would have become wider ranging. A portable kit, like Mode 2, though more costly would have been increasingly advantageous. It would have been the product of increased innovation under increasing stress. Why did Mode 2 not evolve, then, among the Eurasian populations that would have been under even greater stress? These Eurasian Mode 1 populations would have been at much lower densities so it is unlikely that they would have reached the densities that created the kinds of demographic pressure that affected the African populations. The low density would have statistically reduced the chances of innovations appearing and spreading. So, in Eurasia, we would expect thinly spread populations continuing with Mode 1.

As the world became increasingly unstable and all populations adopted a more highly mobile form of living, Mode 1 became extinct and Mode 2 survived. The rapid spread of Mode 2 in Europe in an evolved form, after a significant delay in arrival, is perhaps the first example of the ecological release followed by rapid spread that, as we have seen, characterised later human populations. The delay in arrival in Europe, in spite of the greater mobility and dispersal capacity of these populations may reflect worsening environmental conditions that slowed down the rate of spread. We can thus envisage a situation, starting with the bearers of Mode 1, in which dispersal out from Africa is achieved under relatively benign conditions by slow dispersing hominids. Increasingly harsh conditions would have slowed the rate of dispersal of later populations but this would have been counteracted by increased dispersal capacities and more versatile technology. The result is that hominids kept dispersing out from Africa. A further prediction of this model is that, as dispersal ability and technological versatility improved, populations dispersing from Africa should have been increasingly able to expand the range further than previous dispersing populations, turning former barriers into corridors. This prediction would appear to be met. Bearers of Mode 1 dispersed across tropical and sub-tropical regions of Eurasia including the Mediterranean. Bearers of Mode 2 reached areas of temperate western Europe. Bearers of Mode 3 expanded further into areas of Eurasia. Bearers of Mode 4 were the first to colonise the Eurasian Plain and reach North America. I am not making any distinction as to the biological identity of the bearers of these technologies nor indeed whether the different modes dispersed with populations or simply spread through cultural transmission. The net effect is the same – bearers of Mode 4 colonised the largest areas of the planet and bearers of Mode 1 the smallest.

The sobering conclusion that I derive is that it has been the peripheral populations, those in some way weaker and displaced by density-dependent processes, that have adapted to situations of increased risk. As the world has become an increasingly risky place so these populations have been at an advantage and have spread while the less versatile core populations have lost out. This need not be the result of competitive processes. It is far more likely that less versatile populations were reduced and eliminated by instability and others, better able to deal with instability, moved in then to become core populations in their own right, with a slowed rate of change. New peripheral populations then went through a process of increasing adaptive versatility and the process was repeated all over again. The last cases would have included the extinction of the so-called 'archaic' populations (including the Neanderthals), starting as on all previous occasions with those in tropical Africa, and the filling of the empty space left by them by the peripheral Moderns.

Most discussions of the relationship between Moderns and the Neanderthals end with the Neanderthal extinction that is put at around 30–28 kyr in the last sites (Chapter 7). Clearly Neanderthals must have survived somewhat later as these dates are for sites in which the Neanderthal population density must have been high enough for the archaeological record to detect it. It is naïve to think of these as the latest dates of Neanderthal survival. Even so, we can agree that the last Neanderthal populations probably became extinct before the onset of the LGM, at some point towards the end of OIS 3. So, at the onset of the LGM we have a world that is apparently, in Eurasia at least, inhabited only by Moderns. So what happened to these Moderns?

The last glacial maximum

I will focus this discussion on the European and Middle Eastern picture as it is the best documented. With the onset of the LGM and the advance of the ice sheets (Chapter 6) much of Europe was inhospitable to humans (Housley *et al.*, 1997; Bocquet-Appel & Demars, 2000a). We observe a pattern of human dispersion that is not dissimilar to that of earlier cold events, except that on this occasion it is Moderns and not Neanderthals that are involved. The heightening of the

LGM sees the intrusion of temperate environments well into the Mediterranean refugia, most so in the continental peninsulas of the Balkans and Italy, and also across much of the central Iberian tablelands. Steppe dominates these plains and montane vegetation descends even in the thermo-Mediterranean coastal belt. Steppic species reach far south, with mammoth reaching the latitude of Granada in Spain (37° N) and woolly rhinoceros, reindeer and wolverine reaching well into central Iberia (Garcia & Arsuaga, 2003). The worsening of conditions at the end of OIS 3 saw the corresponding expansion of plains adapted humans across the Eurasian Plain (Chapter 7), first with Aurignacian technology and then with Gravettian technology. Towards the start of the LGM these populations became highly dispersed and fragmented in the Eurasian Plains and many local populations would have become extinct. Instead there is a population demographic explosion in the southern refugia, at least within the Iberian Peninsula and parts of south-western France. This increase is related to people carrying a new technology (the Solutrean) that appears adapted to the exploitation of open vegetation plains fauna with long-range weapons (Straus *et al.*, 2000). The increase corresponds to the expansion of steppic environments in Iberia. Some authors have connected this population increase with the decline in the Eurasian Plain, suggesting southward displacement of these populations. Although there may have been an element of southward dispersal as populations retreated, I think that it is more likely that these populations simply were unable to adapt fast enough to the change and went extinct. At the same time those that had penetrated Iberia were adapted to steppe and their populations expanded. This scenario is not dissimilar to those that Neanderthals would have faced in previous glaciations with one exception. For the first time we have steppe-adapted populations of humans within the refugia able to capitalise on the spread of these habitats. But we saw in Chapter 5 that one difference between Moderns and Neanderthals was that the former could also exploit heterogeneous environments. So the people of the Solutrean were also able to survive along the coastal mountains of Cantabria and the Mediterranean coast of Iberia.

The last deglaciation

The deglaciation was marked by a demographic explosion (Demars, 1996; Bocquet-Appel & Demars, 2000a) and by the expansion of humans from southern and eastern refugia into the Eurasian Plains as the ice sheets receded (Aldhouse-Green & Pettitt, 1998; Torroni *et al.*, 1998, 2001; Street & Terberger, 1999; Blockley *et al.*, 2000; Semino *et al.*, 2000). The size of Iberia meant that it probably held the largest population of humans in Europe, so it is not surprising to see the extent of their subsequent genetic influence as they spread northwards (Torroni *et al.*, 1998, 2001; Semino *et al.*, 2000). The Cantabrian populations of Moderns expanded further and colonised upland areas as these opened up with global warming (Straus *et al.*, 2000). This is not surprising as these populations were experiencing similar Euro-Siberian vegetation and faunal conditions to those further north. The Mediterranean ones were, instead, on the decline (Figure 7.8; Straus *et al.*, 2000). My view is that as the Mediterranean forests expanded many previously suitable areas of steppe or heterogeneous landscapes became unavailable. Forests were challenging on account of the dispersed nature of prey species, that would also have been less visible, and the difficulty of penetration into their depths. So the southern populations were restricted to the coastal areas where many, perhaps resembling the response of earlier populations of Neanderthals under similar circumstances, became heavily dependent on marine resources. This dispersion pattern within Iberia continued into the early Holocene, Epi-Palaeolithic populations being largely distributed coastally in the Mediterranean (Oliver & Juan-Cabanilles, 2000) and in Portugal and extending into high mountains in the north (Straus *et al.*, 2000).

In the Eurasian Plain we see the cultural and technological explosion that we relate to the Magdalenian. Following from my earlier point about the development of innovations in times of stress, we observe the beginnings of this diversification in the Solutrean, which corresponds to the period of greatest population stress. The Magdalenian is the first great cultural and technological diversification of the Modern Humans, breaking away from dependence on mammalian herbivore meat towards the exploitation of a diversity of resources (Chapter 5; Holliday, 1998). These changes were themselves induced by environmental stresses as the mammalian herbivore fauna of the Eurasian Plain, on which humans had become dependent since the Aurignacian, became depleted. Some authors consider this depletion to have been induced by the Palaeolithic hunters themselves but others prefer to regard the late Pleistocene megaherbivore extinction to have been induced by climate driven habitat change. Either way, Modern Humans had to, for the first time, develop lifestyles that were not heavily dependent on the hunting of large mammalian herbivores. The large packages of protein (Chapter 2) that had characterised the landscapes in which humans evolved were significantly reduced.

The deglaciation and the Holocene were not the first period of global warming that humans experienced in Europe and western Asia. Neanderthals recovered from earlier glacials as we saw in Chapter 7. However, they did not diversify anywhere near to the extent that the Magdalenian Moderns did. The reason may simply be due to the cumulative effects that I have alluded to that led to the progressive increase in technological complexity that we observe in the archaeological record. This does not mean that the Neanderthals could not have diversified in the way that the Magdalenians did. It simply means that the starting template of the Neanderthals and that of the Magdalenians was different, largely due to reasons of contingency. Additionally, the faunal pattern would have been very different in the two cases. The Neanderthals would have lived in a world that remained rich in mammalian herbivores, whereas the Magdalenians, as we have seen, lived in a world of progressive impoverishment of mammalian herbivore diversity and biomass. Nevertheless, we have tantalising glimpses that suggest that Neanderthals may have responded in a similar way to Magdalenians with regard to resource exploitation. During OIS 5, at Vanguard Cave, Gibraltar, Neanderthals were exploiting marine resources including seals and dolphins (Balter, 2001; Chapter 5). It suggests the possibility that, as forests covered Iberia and the continental shelf was submerged as a result of global warming, the southern Iberian Neanderthals at least were diversifying their resource base. We cannot be certain of this and must await further evidence of Neanderthal behavioural ecology during the last interglacial.

Systems of food production

Why did the Neanderthals not evolve food production during the last inter-glacial? Why did the early Moderns not do so either? Indeed, some societies – such as the aboriginal Australians – never acquired food production at all (Diamond, 1997). These may seem pedantic questions but attempting to an-swer them may help us understand the conditions in which food production might be expected to arise. We have seen that Neanderthals differed from Mod-erns in the way in which they exploited the landscape, frequently moving within a relatively small annual home range. We have no evidence that Neanderthals cached resources and it appears unlikely that they remained for long periods at any particular location. I have also suggested that there was little, if any, divi-sion of labour in these Neanderthal groups that were, in all probability, small. A highly mobile strategy largely based on the exploitation by encounter of re-sources that were highly dispersed would not seem suited for the conditions required for agriculture (semi-permanent or sedentary existence, within-group division of labour) to evolve. We therefore do not need to seek cognitive ca-pacity explanations (e.g. Diamond, 2002) to explain why food production only evolved in Moderns. The early Moderns were evolving a type of mobility pat-tern that might have provided suitable conditions for such a development to take place during the last interglacial, but it never happened. It seems that these early Moderns were simply not in the right place at the right time. During the last interglacial they were still confined to tropical Africa. This gives us another insight – the geographical component in the evolution of agriculture.

Agriculture has originated independently on several occasions. The geographical pattern shows that agriculture has never originated in areas of the Eurasian Plain. The reason for this is a combination of absence of the plants suitable for domestication, the seasonality related to latitude and the harshness of the winters. Indeed, reindeer herding has been the only form of food production that has evolved in the Eurasian Arctic (Diamond, 1997). Agriculture has rarely originated in the tropics either, although there are examples in, for example, Mesoamerica, the Andes and Amazonia (Diamond, 1997). The reason for this may be harder to explain. There are suitable plants, for example maize, beans and squash in Mesoamerica and potato and manioc in the Andes and Amazonia, which were widely domesticated. It is also hard to find satisfactory climatic explanations. The reason may lie in the people themselves. Most humans living in the tropics did not experience the same level of shortages of mammalian herbivores that those in the temperate regions did. In addition, tropical regions provided a wider range of food resources that could be gathered or hunted than temperate areas. So it seems that the pressures to develop food production were simply not present.

The significant early Holocene developments in agriculture have taken place in areas across the Eurasian mid-latitude belt – the fertile crescent, and China in particular (Diamond, 1997; Lev-Yadun, 2000). From these nuclei agriculture spread to adjacent areas with similar climatic characteristics. These areas held a range of plant species that were suitable for domestication (Diamond, 1997). Climatic conditions were such that mammalian herbivores, after the late Pleistocene extinctions, were scarcer than in tropical areas though they were not wholly unavailable. The diversity of alternative food resources would have been lower than in the tropics. Yet climate was sufficiently benign to permit the sustainable development of agriculture. Even so, foraging lifestyles continued in areas of, for example, the Iberian Peninsula where the rich resources of estuaries allowed this even after the arrival of Neolithic food production systems (Straus *et al.*, 2000) brought by maritime pioneers around 7.5 kyr (Zilhao, 2001). The rich marine resources of the Strait of Gibraltar also permitted a subsistence based on the harvesting of marine resources in the Neolithic (Finlayson *et al.*, 1999). In Japan, hunting–fishing–gathering societies persisted until 300 BC when the transition to rice cultivation took place (Akazawa, 1996b). Exploitation of marine mammals off the Californian coast from 7040 BC to AD 1400 indicates continued use of widely available wild resources (Porcasi *et al.*, 2000). Where available, and with the development of appropriate sea-going technology, marine mammals became viable alternatives to terrestrial mammalian herbivores.

All this indicates pressures towards retention of hunter–gatherer lifestyles whenever these were sustainable. However, once populations went down the route of food production, the competitive advantages conferred on farmers (in spite of the many disadvantages, e.g. worsened nutritional condition and greater propensity to disease) made the transition from hunting and gathering to food production autocatalytic (Diamond, 2002). Among the advantages of a sedentary, food production, lifestyle is the reduction of the birth–birth interval with its consequences for population growth (Diamond, 1997, 2002). Here we see the outcome of a process of birth–birth interval reduction that would have started with the social behaviour and mobility patterns of the early Moderns (Chapter 5) that would have enabled a more rapid population growth and reduced risk of extinction, all else being equal, than would have been the case with the Neanderthals (Zubrow, 1989).

So what can we conclude were the essential characteristics required for the origin of agriculture in the early Holocene? There were seven prerequisites: (1) the availability of plants suitable for domestication (Diamond, 1997); (2) wild resources permitting a hunting and gathering existence but with seasonal deficits and an insufficient range of alternatives to see populations through the bad times; (3) a climate suitable for the growth of domesticated plants; (4) heterogeneous landscapes that were characterised by high ecological diversity and stepwise seasonal climatic shifts that would have favoured staggered growth of crops (Diamond, 1997) and that would also have permitted an element of hunting and gathering; (5) populations of humans with territorial exploitation strategies that required the semi-permanent or sedentary existence of at least part of the population; (6) populations of humans that had well developed technology, systems of division of labour and storage; and (7) ecologically stressed populations of humans.

Clearly, the climatic amelioration of the Holocene linked with the Modern Human lifestyles that had evolved in the late Pleistocene increased the probability of some groups of humans discovering the benefits of particular plants for domestication. These would have originally grown naturally around camp sites. Without an element of residence the establishment of the link between the seasonal re-emergence of particular plants and the collection of their seeds would be highly improbable. The pressure towards sedentism may have been demographic. Sedentism would have evolved to ensure access to the most productive portions of a territory in the context of increasing competition (Rosenberg, 1998). Innovations would have arisen, as throughout human history, among stressed populations living outside the tropics. Those in the temperate and boreal regions were unable to develop agriculture because of unsuitable plants and climatic conditions. Those across the mid-latitude belt were prime candidates. That agriculture arose independently on at least two occasions between 10.5 and 9.5 kyr, when it had not done so previously in the history of human evolution, suggests that it could only have done so when all seven factors listed above coincided. The later independent origins of agriculture in other parts of the world also took place when all seven factors coincided.

Similar factors would have contributed to the herding, and subsequent domestication, of mammalian herbivores. Herding would have evolved as an extended form of storage, keeping the animals alive for use during the lean periods of the year. The selection of those animals with docile characteristics would have progressively generated strains that were easy to keep around settlements. This is exactly what appears to have occurred in the case of goat domestication in the Zagros Mountains in western Iran (Zeder & Hesse, 2000). Here the domestication process appears to have started around 10 kyr (and possibly even earlier) through herd management. During the early phases managed goats may have interbred freely with wild ones and selective breeding and control of sex ratios and age structures was only gradually introduced afterwards.

I described in Chapter 2 how humans preferentially hunted intermediate size mammals throughout their history. The preference for these intermediate size classes continued to be exercised in the case of domesticated mammals. Successfully domesticated large mammals, those that have been the most important to humans, were almost exclusively Eurasian in origin (Diamond, 1997). Most of the large mammals that were domesticated, including four of the five main ones – goat, sheep, pig and cow, had geographical distributions that included the mid-latitude belt so it is not surprising to find early evidence of domestication there, particularly in western Asia. Domestication was not limited, however, to this belt. The fifth species, the horse, was essentially an animal of the steppes that was herded and then domesticated in the plains of central Asia. Diamond (1997) has suggested that the absence of native mammal domestication in Africa, the Americas and Australia was related to the pool of locally available species that could be domesticated. Although this may be true it may only be part of the explanation. Hunting–gathering probably remained viable in most of these regions, indeed continued until today in Australia, so that the pressure to move towards animal (and indeed plant) domestication was not as great as in the mid-latitude belt or the plains of Eurasia. Mid-latitude Eurasia continued to meet the seven criteria outlined above for the origins of agriculture. The Eurasian Plains satisfied these in respect of the horse. Horse domestication was in any case initiated, over 7 kyr, by people who also kept cattle and sheep (Anthony & Brown, 2000) so that the knowledge of herding and domestication was probably imported and applied to a locally-abundant species. The domestication of the horse may appear not to fulfil criterion (4) below. The plains permitted a wide-ranging strategy and movement between pastures at a scale that would not have been possible for species in the heterogeneous belt. The consequences would have been similar to those for the smaller-scale seasonal movements of, for example, goats and sheep – animals could be moved to the best grazing areas at any particular point. The subsequent development of means of storing fodder for the lean seasons was a cultural way of providing for these needs. Horse domestication was also exceptional in that its main function was in increasing mobility, at least from 5.5 kyr, even though it had started as a means of providing an additional component in the diet (Levine, 1999; Anthony & Brown, 2000). We have seen how, since the colonisation of the Eurasian Plains by Moderns, mobility across their vast spaces that were largely free of topographic obstacles was the key to survival. I view the domestication of the horse as a natural extension of this adaptation, one that was to have significant consequences in the later historical evolution of Eurasia (Fernandez-Armesto, 2000).

We can therefore refine our seven criteria to incorporate animal domestication: (1) the availability of organisms suitable for domestication; (2) wild resources permitting a hunting and gathering existence but with seasonal deficits and an insufficient range of alternatives to see populations through the bad times; (3) a climate suitable for the growth or management of domesticated organisms; (4) heterogeneous landscapes that were characterised by high ecological diversity and stepwise seasonal climatic shifts that would have favoured staggered growth of crops or seasonal movement of herding animals to new pastures and that would also have permitted an element of hunting and gathering; (5) populations of humans with territorial exploitation strategies that required the semi-permanent or sedentary existence of at least part of the population; (6) populations of humans that had well developed technology, systems of division of labour and storage; and (7) ecologically stressed populations of humans.

Two alternative ways of being human

We have seen how the unpredictable events of the Pleistocene were the driving force behind human evolution. The colonisation of Europe and western Asia by a Middle Eastern or, most probably, African population perhaps as far back as half-a-million years ago led to a lineage of humans that evolved in relative isolation. This lineage became identifiable as the 'classic' Neanderthals around 120 kyr. The Neanderthals were intelligent humans, derived from the common stock that also led to the Modern Human lineage. Their morphology was adapted to those features that were most stable in the mid-latitude belt that stretched from Portugal to the central Asian mountains – the heterogeneous, species rich, landscapes of the lower parts of the mountain ranges that dominated this vast territory and the large mammalian herbivores that they consumed. Their intelligence permitted a degree of behavioural flexibility that enabled them to survive through harsh as well as mild climatic conditions. The system permitted their survival for tens of thousands of years. The constant Neanderthal populations were distributed across the mid-latitude belt. Those that colonised more temperate areas during mild climatic conditions became extinct each time harsh conditions returned to Europe. The frequency of these harsh conditions increased during the last glacial cycle and especially towards the end of OIS 3 around 50 kyr. Neanderthal populations became increasingly stressed. The last populations, in a typically human response, attempted to adapt behaviourally through technological innovation and changes in mobility but their morphology, that had served them so well, could not change fast enough to permit them to deal with the long range movements of the herbivores that increasingly dominated a new world dominated by treeless vegetation. The last Neanderthal populations became fragmented, isolated from each other and they disappeared one after the other over a relatively short time period between 40 and 25 kyr. The causes of the extinction of each population need not have been the same. They may have included local competition from Modern Human groups, Allee effects, disease, inbreeding, even random population fluctuations. An intelligent and alternative way of being human became extinct.

A second intelligent type of human, closely related to the Neanderthals having only diverged from them half-a-million years earlier, that had increasingly invested in a gracile morphology that permitted them to reduce the risks of the spatial and short-term temporal unpredictability of the African plains by increasing their scale of operation, emerged from Africa around 100 kyr. These Moderns had gradually developed increasingly complex and symbolic social behaviours that improved the flow of information between members of large groups that exploited large home ranges. Large group sizes, home ranges and complex social behaviour together enabled these Moderns to minimise risks of living in unpredictable environments. The development of the neocortex in these humans provided the necessary hardware for dealing with the complex interactions within the group.

These early Moderns were apparently unable to penetrate immediately onto the Eurasian Plain. It is very possible that the presence of the highly adapted Neanderthals kept them in the periphery. They managed to spread eastwards, perhaps via the Horn of Africa, keeping within tropical and sub-tropical climates and probably reached Australia by at least 60 – 55 kyr judging from the presence of Moderns in Lake Mungo, deep in Australia, by 50 kyr. By 45 kyr, Neanderthal populations were on the decline and this may have opened a window for the Moderns to penetrate the Eurasian Plain. They may have done so simultaneously along various points, into the Balkans from Asia Minor and onto the Russian Plain from the Caucasus. I prefer the second alternative which has some support from genetics and archaeology.

The complex behaviours of the Moderns became even more accentuated and diversified on the plains of Eurasia. The rich mammalian herbivore resources scattered across vast open landscapes required complex ways of locating them, accurate information transfer systems between individuals in the group, portable and projectile technology, and social binding behaviours. Symbolism became particularly important in a world in which individuals, for the first time, were required to transmit and receive information relating to events that others had not directly experienced.

These behaviours were not potentially unique to the Moderns. The Neanderthals could have evolved similar behaviours had they been exposed to the social and ecological pressures demanded by this alternative way of dealing with the world. I therefore maintain that Neanderthals were not cognitively inferior to Moderns nor do I find support for a sudden mutation that generated the complete package that made up the Modern Human overnight. Neanderthals were simply different. A morphology and way of life that had become increasingly complex and had been successful across the mid-latitude belt of western and central Eurasia for close to half-a-million years disappeared because it was not designed to cope with the speed and direction of change that hit Eurasia at the end of the Pleistocene. The Moderns managed to survive the onslaught of the LGM. With the subsequent global warming the world saw the rise of a hominid that had the intelligence and an accumulated tradition of increasingly complex technology, social behaviour that enabled refined interactions within large groups, systems of storage that reduced risk in times of shortage and mobility strategies that together provided, in the right geographical circumstances, the necessary templates to deal with future stresses. These stresses came from an expanding population and a reducing resource base. In the Middle East marginal groups that were finding a hunting and gathering existence increasingly difficult learnt new ways of dealing with uncertainty, a trend that had started with the development of stone tools two million years earlier. They had the necessary behavioural templates that allowed them to herd animals, domesticate them, plant the seeds of wild plants and improve them.

Moderns have continued the trends towards increasingly complex technologies and social systems, having conquered even the most inhospitable of environments (Gamble, 1993; Fernandez-Armesto, 2000). Cultural and social diversity is the hallmark of human societies across the Earth today yet nobody seriously attempts to equate these differences to biological differences. Nobody, rightly, suggests that we are observing different species of humans. Yet, looking at similar evidence of cultural and social diversity in the Pleistocene there are still those who equate these to biological differences, the product of mutations that made us something apart from the rest of nature. It is just another version of the antiquated view of humans at the top of the evolutionary pyramid. If anything, I hope to have shown in this book that we are the product of chance and a great deal of luck. We are here because, in scrambling for survival in the margins of the world of other humans, we became increasingly inventive and kept finding ways of hanging on and then taking over when others that had been better adapted than ourselves vanished as circumstances changed. That we are here today is the end result of a series of chance events that kept us in the running. It could easily have gone the other way . . .

References

Abegg, C. & Thierry, B. (2002). Macaque evolution and dispersal in insular south-east Asia. *Biological Journal of the Linnean Society*, **75**, 555–576.

Adams, J. M. & Faure, H. (1997). Preliminary vegetation maps of the world since the Last Glacial Maximum: an aid to archaeological understanding. *Journal of Archaeological Science*, **24**, 623–647.

Adcock, G. J., Dennis, E. S., Easteal, S. *et al.* (2001). Mitochondrial DNA sequences in ancient Australians: implications for modern human origins. *Proceedings of the National Academy of Sciences, USA*, **98**, 537–542.

Aguirre, E. (1994). *Homo erectus* and *Homo sapiens*: one or more species? *Courier Forschungs-Institut Senckenberg*, **171**, 333–339.

Aguirre, E. (2000). Poor fossil record and major changes around 1 MaBP. *Human Evolution*, **15**, 51–62.

Aguirre, E. & Carbonell, E. (2001). Early human expansions into Eurasia: The Atapuerca evidence. *Quaternary International*, **75**, 11–18.

Aiello, L. C. & Dean, M. C. (1990). *An Introduction to Human Evolutionary Anatomy*. London: Academic Press.

Aiello, L. C. & Dunbar, R. I. M. (1993). Neocortex size, group size, and the evolution of language. *Current Anthropology*, **34**, 184–193.

Aiello, L. C. & Wheeler, P. (1995). The expensive-tissue hypothesis: the brain and the digestive system in human and primate evolution. *Current Anthropology*, **36**, 199–221.

Akazawa, T. (1996a). Introduction: human evolution, dispersals, and adaptive strategies. In *Prehistoric Mongoloid Dispersals*, ed. T. Akazawa & E. J. E. Szathmary, pp. 1–37. Oxford, Oxford University Press.

Akazawa, Y. (1996b). Regional variation in Jomon hunting–fishing– gathering societies. In *Interdisciplinary Perspectives on the Origins of the Japanese*, pp. 223–231. Tokyo, International Research Center for Japanese Studies.

Aldhouse-Green, S. & Pettitt, P. (1998). Paviland Cave: contextualizing the 'Red Lady'. *Antiquity*, **72**, 756–772.

Alimen, H. (1975). Les "isthmes" hispano-marocain et siculo-tunisien aux temps acheuleens. *L'Anthropologie*, **79**, 399–436.

Allen, J. R. M., Brandt, U., Brauer, A. *et al.* (1999). Rapid environmental changes in southern Europe during the last glacial period. *Nature*, **400**, 740–743.

Alley, R. B. (2000). Ice-core evidence of abrupt climate changes. *Proceedings of the National Academy of Sciences, USA*, **97**, 1331–1334.

Alley, R. B., Lynch-Stieglitz, J. & Severinghaus, J. P. (1999). Global climate change. *Proceedings of the National Academy of Sciences, USA*, **96**, 9987–9988.

Allsworth-Jones, P. (1993). The archaeology of archaic and early modern *Homo sapiens*: an african perspective. *Cambridge Archaeological Journal*, **3**, 21–39.

Alonso, S. & Armour, J. A. L. (2001). A highly variable segment of human subterminal 16p reveals a history of population growth for modern humans outside Africa. *Proceedings of the National Academy of Sciences, USA*, **98**, 864–869.

Alroy, J. (1998). Cope's rule and the dynamics of body mass evolution in North American fossil mammals. *Science*, **280**, 731–734.

Ambrose, S. H. (1998). Late Pleistocene human population bottlenecks, volcanic winter, and differentiation of modern humans. *Journal of Human Evolution*, **34**, 623–651.

Ambrose, S. H. (2001). Paleolithic technology and human evolution. *Science*, **291**, 1748–1753.

Anderson, A. (2002). Faunal collapse, landscape change and settlement history in Remote Oceania. *World Archaeology*, **33**, 375–390.

Anderson, C. M. (1989). Neandertal pelves and gestation length: hypotheses and holism in paleoanthropology. *American Anthropologist*, **91**, 327–340.

Anderson, C. & Franks, N. R. (2001). Teams in animal societies. *Behavioral Ecology*, **12**, 534–540.

Anderson, E. (1984). Who's who in the Pleistocene: a mammalian bestiary. In *Quaternary Extinctions. A Prehistoric Revolution*, ed. P. S. Martin & R. G. Klein, pp. 40–89. Tucson, University of Arizona Press.

Andreev, A. A., Siegert, C., Klimanov, V. A., Derevyagin, A. Y., Shilova, G. N. & Melles, M. (2002). Late Pleistocene and Holocene vegetation and climate on the Taymyr Lowland, Northern Siberia. *Quaternary Research*, **57**, 138–150.

Anthony, D. W. & Brown, D. R. (2000). Eneolithic horse exploitation in the Eurasian steppes: diet, ritual and riding. *Antiquity*, **74**, 75–86.

Anton, S. C., Leonard, W. R. & Robertson, M. L. (2002). An ecomorphological model of the initial hominid dispersal from Africa. *Journal of Human Evolution*, **43**, 773–785.

Antunes, M. T., Cabral, J. M. P., Cardoso, J. L., Pais, J. & Soares, A. M. (1989). Paleolítico médio e superior em Portugal: datas ^{14}C, estado actual dos conhecimentos, sintese e discussao. *Ciencias da Terra*, **10**, 127–138.

Arensburg, B. & Belfer-Cohen, A. (1998). Sapiens and Neandertals: rethinking the Levantine middle Paleolithic hominids. In *Neandertals and Modern Humans in Western Asia*, ed. T. Akazawa, K. Aochi & O. Bar-Yosef, pp. 311–338. New York, Plenum Press.

Arnold, N. S., van Andel, T. H. & Valen, V. (2002). Extent and Dynamics of the Scandinavian Ice Sheet during Oxygen Isotope Stage 3 (65,000–25,000 yr BP). *Quaternary Research*, **57**, 38–48.

Arribas, A. & Palmqvist, P. (1999). On the ecological connection between sabre-tooths and hominids: faunal dispersal events in the lower Pleistocene and a review of the evidence for the first human arrival in Europe. *Journal of Archaeological Science*, **26**, 571–585.

Arsuaga, J. L., Martínez, I., Gracia, A., Carretero, J. M. & Carbonell, E. (1993). Three new human skulls from the Sima de los Huesos middle Pleistocene site (Atapuerca, Spain). *Nature*, **362**, 534–537.

Arsuaga, J. L., Martínez, I., Gracia, A. & Lorenzo, C. (1997). The Sima de los Huesos crania (Sierra de Atapuerca, Spain): a comparative study. *Journal of Human Evolution*, **33**, 219–281.

Asfaw, B., Henry Gilbert, W., Beyenne, Y. *et al.* (2002). Remains of *Homo erectus* from Bouri, Middle Awash, Ethiopia. *Nature*, **416**, 317–320.

Aura Tortosa, J. E. (1989). Solutrenses y Magdalenienses al sur del Ebro. Primeras aproximaciones a un proceso de cambio industrial: el ejemplo de Parpalló. *Plav-Saguntum*, **22**, 35–65.

Awadalla, P., Eyre-Walker, A. & Maynard Smith, J. (1999). Linkage disequilibrium and recombination in hominid mitochondrial DNA. *Science*, **286**, 2524–2525.

Axtell, R. L., Epstein, J. M., Dean, J. S. *et al.* (2002). Population growth and collapse in a multiagent model of the Kayenta Anasazi in Long House Valley. *Proceedings of the National Academy of Sciences, USA*, **99**, 7275–7279.

Balter, M. (2001). What or who did in the Neandertals? *Science*, **293**, 1980–1981.

Barbujani, G. & Bertorelle, G. (2001). Genetics and the population history of Europe. *Proceedings of the National Academy of Sciences, USA*, **98**, 22–25.

Barham, L. S. (1998). Possible early pigment use in south-central Africa. *Current Anthropology*, **39**, 703–710.

Barham, L. (2002a). Backed tools in Middle Pleistocene central Africa and their evolutionary significance. *Journal of Human Evolution*, **43**, 585–603.

Barham, L. S. (2002b). Systematic pigment use in the Middle Pleistocene of south-central Africa. *Current Anthropology*, **43**, 181–190.

Barlow, L. K., Sadler, J. P., Ogilvie, A. *et al.* (1998). Ice core and environmental evidence for the end of Norse Greenland. *The Holocene*, **7**, 489–499.

Bar-Matthews, M., Ayalon, A. & Kaufman, A. (1997). Late Quaternary Paleoclimate in the Eastern Mediterranean Region from Stable Isotope Analysis of Speleothems at Sorqe Cave, Israel. *Quaternary Research*, **47**, 155–168.

Barron, E. & Pollard, D. (2002). High-resolution climate simulations of Oxygen Isotope Stage 3 in Europe. *Quaternary Research*, **58**, 296–309.

Barton, N. H. & Hewitt, G. M. (1985). Analysis of hybrid zones. *Annual Review of Ecology and Systematics*, **16**, 113–148.

Bar-Yosef, O. (1994). The contribution of south-west Asia to the study of modern human origins. In *Origins of Anatomically Modern Humans*, ed. M. H. Nitecki & D. V. Nitecki, pp. 22–36. New York, Plenum Press.

Bar-Yosef, O. (1996). The impact of late Pleistocene-early Holocene climatic changes on humans in southwest Asia. In *Humans at the End of the Ice Age*, ed. L. G. Straus, B. V. Eriksen, J. M. Erlandson & D. R. Yesner, pp. 61–78. New York: Plenum Press.

Bar-Yosef, O. (1998). The Chronology of the Middle Paleolithic of the Levant. In *Neandertals and Modern Humans in Western Asia*, ed. T. Akazawa, K. Aochi & O. Bar-Yosef, pp. 39–56. New York, Plenum Press.

Bar-Yosef, O. (2000). The impact of radiocarbon dating on Old World archaeology: past achievements and future expectations. *Radiocarbon*, **42**, 23–39.

Bar-Yosef, O. & Belfer-Cohen, A. (2001). From Africa to Eurasia – early dispersals. *Quaternary International*, **75**, 19–28.

Bar-Yosef, O. & Kuhn, S. (1999). The big deal about blades: laminar technologies and human evolution. *American Anthropology*, **101**, 1–17.

Bazzaz, F. A. (1996). *Plants in Changing Environments. Linking physiological, population, and community ecology*. Cambridge, Cambridge University Press.

Beals, K. L., Smith, C. L. & Dodd, S. M. (1984). Brain Size, Cranial Morphology, Climate, and Time Machines. *Current Anthropology*, **25**, 301–330.

Bell, S. S., McCoy, E. D. & Mushinsky, H. R., eds. (1991). *Habitat Structure. The Physical Arrangement of Objects in Space*. London, Chapman & Hall.

Belmaker, M., Tchernov, E., Condemi, S. & Bar-Yosef, O. (2002). New evidence for hominid presence in the Lower Pleistocene of the Southern Levant. *Journal of Human Evolution*, **43**, 43–56.

Bennett, K. D., Tzedakis, P. C. & Willis, K. J. (1991). Quaternary refugia of north European trees. *Journal of Biogeography*, **18**, 103–115.

Berggren, W. A., Kent, D. V., Swisher, C. C. & Aubry, M. P. (1995). A revised Cenozoic geochronology and chronostratigraphy. In *Geochronology, Time Scales, and Global Stratigraphic Correlations: A Unified Temporal Framework for A Historical Geology*, ed. W. A. Berggren, D. V., Kent, & J. Hardenbol. Society for Economic Paleontology & Mineralogy, Special Publication, **54**: 129–212. Tulsa, Society for Economic Paleontology & Mineralogy.

Bermúdez de Castro, J. M., Arsuaga, J. L., Carbonell, E., Rosas, A., Martínez, I. & Mosquera, M. (1997). A hominid from the Lower Pleistocene of Atapuerca, Spain: possible ancestor to Neandertals and modern humans. *Science*, **276**, 1392–1395.

Bigalke, R. C. (1968). The contemporary mammal fauna of Africa. *Quarterly Review of Biology*, **43**, 265–300.

Binford, L. R. (1980). Willow smoke and dogs' tails: hunter–gatherer settlement systems and archaeological site formation. *American Antiquity*, **45**, 4–20.

Binford, M. W., Kolata, A. L., Brenner, M. *et al.* (1997). Climate Variation and the Rise and Fall of an Andean Civilization. *Quaternary Research*, **47**, 235–248.

Birks, H. J. B. (1986). Late Quaternary biotic changes in terrestrial and lacustrine environments, with particular reference to north-west Europe. In *Handbook of Holocene Palaeoecology and Palaeohydrology*, ed. B. E. Berglund, pp. 3–65. Chichester, John Wiley & Sons.

Bishop, L. C. (1999). Suid Paleoecology and habitat preferences at African Pliocene and Pleistocene hominid localities. In *African Biogeography, Climate Change, and Human Evolution*, ed. T. G. Bromage & F. Schrenk, pp. 216–225. Oxford, Oxford University Press.

Blades, B. (1999). Aurignacian Settlement Patterns in the Vezere Valley. *Current Anthropology*, **40**, 712–718.

Blasi, C., Carranza, M. L., Filesi, L., Tilia, A. & Acosta, A. (1999). Relation between climate and vegetation along a Mediterranean-Temperate boundary in central Italy. *Global Ecology and Biogeography*, **8**, 17–27.

Bleed, P. (1986). The optimal design of hunting weapons: maintainability or reliability. *American Antiquity*, **51**, 737–747.

Blockley, S. P. E., Donahue, R. E. & Pollard, A. M. (2000). Radiocarbon calibration and Late Glacial occupation of northwest Europe. *Antiquity*, **74**, 112–119.

Blondel, J. & Aronson, J. (1999). *Biology and Wildlife of the Mediterranean Region*. Oxford, Oxford University Press.

Bobe, R., Behrensmeyer, A. K. & Chapman, R. E. (2002). Faunal change, environmental variability and late Pliocene hominin evolution. *Journal of Human Evolution*, **42**, 475–497.

Bocherens, H., Fizet, M., Mariotti, A. *et al.* (1991). Isotopic biogeochemistry (^{13}C, ^{15}N) of fossil vertebrate collagen: application to the study of a past food web including Neandertal man. *Journal of Human Evolution*, **20**, 481–492.

Bocherens, H., Billiou, D., Mariotti, A. *et al.* (1999). Palaeoenvironmental and palaeodietary implications of isotopic biogeochemistry of last interglacial Neanderthal and mammal bones in Scladina Cave (Belgium). *Journal of Archaeological Science*, **26**, 599–607.

Bocquet-Appel, J-P. & Demars, P-Y. (2000a). Population kinetics in the Upper Palaeolithic in Western Europe. *Journal of Archaeological Science*, **27**, 551–570.

Bocquet-Appel, J-P. & Demars, P. Y. (2000b). Neanderthal contraction and modern human colonization of Europe. *Antiquity*, **74**, 544–552.

Boettger, T., Junge, F. W. & Litt, T. (2000). Stable climatic conditions in central Germany during the last interglacial. *Journal of Quaternary Science*, **15**, 469–473.

Bond, G. C. & Lotti, R. (1995). Iceberg discharges into the North Atlantic on millennial time scales during the last glaciation. *Science*, **267**, 1005–1010.

Bond, G., Showers, W., Cheseby, M. *et al.* (1997). A pervasive millennial-scale cycle in North Atlantic Holocene and glacial climates. *Science*, **278**, 1257–1266.

Bordes, F. (1961). Mousterian cultures in France. *Science*, **134**, 803–810.

Bosch, E., Calafell, F., Comas, D., Oefner, P. J., Underhill, P. A. & Bertranpetit, J. (2001). High-resolution analysis of human Y-chromosome variation shows a sharp discontinuity and limited gene flow between northwestern Africa and the Iberian Peninsula. *American Journal of Human Genetics*, **68**, 1019–1029.

Bouzouggar, A., Kozlowski, J. K. & Otte, M. (2002). Étude des ensembles lithiques atériens de la grotte d'El Aliya a Tanger (Maroc). *L'Anthropologie*, **106**, 207–248.

Bowler, J. M., Johnston, H., Olley, J. M. *et al.* (2003). New ages for human occupation and climatic change at Lake Mungo, Australia. *Nature*, **421**, 837–840.

Boyle, K. (1990). *Upper Paleolithic Faunas from South-west France: A Zoogeographic Perspective*. Oxford, British Archaeological Reports Series 557.

Brace, C. L., Ryan, A. S. & Smith, B. D. (1981). Comment on 'Tooth wear in La Ferrassie Man.' *Current Anthropology*, **22**, 426–430.

Brandon-Jones, D. (1996). The Asian Colobinae (Mammalia: Cercopithecidae) as indicators of Quaternary climatic change. *Biological Journal of the Linnean Society*, **59**, 327–350.

Brantingham, P. J., Krivoshapkin, A. I., Jinzeng, L. & Tserendagva, Y. (2001). The initial Upper Palaeolithic in northeast Asia. *Current Anthropology*, **42**, 735–746.

Brauer, G. (1992). Africa's place in the evolution of modern humans. In *Continuity or Replacement. Controversies in* Homo sapiens *Evolution*, ed. G. Brauer & F. H. Smith, pp. 63–98. Rotterdam, A. A. Balkema.

Brauer, G., Yokoyama, Y., Falgueres, C. & Mbua, E. (1997). Modern human origins back-dated. *Nature*, **386**, 337–338.

Broecker, W. (1984). Terminations. In *Milankovitch and Climate*, ed. A. Berger, J. Imbrie, G. Hays, G. Kukla & B. Saltzman. pp. 687–698. Dordrecht, Reidel.

Broecker, W. S. & Hemming, S. (2001). Climate swings come into focus. *Science*, **294**, 2308–2309.

Bromage, T. G. & Schrenk, F. (Eds) (1999). *African Biogeography, Climate Change, and Human Evolution*. Oxford, Oxford University Press.

Brooks, A. S., Helgren, D. M., Cramer, J. S. *et al.* (1995). Dating and context of three Middle Stone Age sites with bone points in the Upper Semliki Valley, Zaire. *Science*, **268**, 548–553.

Brose, D. & Wolpoff, M. (1971). Early Upper Palaeolithic Man and late Middle Palaeolithic tools. *American Anthropologist*, **73**, 1156–1194.

Brown, J. H. (1995). *Macroecology*. Chicago, University of Chicago Press.

Brown, J. H. & Gibson, A. C. (1983). *Biogeography*. St. Louis, C. V. Mosby.

Brown, J. H. & Kodric-Brown, A. (1977). Turnover rates in insular biogeography: effect of immigration on extinction. *Ecology*, **58**, 445–449.

Brunet, M., Guy, F., Pilbeam, D. *et al.* (2002). A new hominid from the Upper Miocene of Chad, Central Africa. *Nature*, **418**, 145–151.

Buckland, P. C., Buckland, P. I. & Skidmore, P. (1998). Insect remains from GUS: an interim report. In *Man, Culture and Environment in Ancient Greenland*, ed. J. Arneborg & H. C. Gullov, pp. 74–79. Copenhagen, Dansk Polar Centre & Danish National Museum.

Buckley, C. & Steele, J. (2002). Evolutionary ecology of spoken language: co-evolutionary hypotheses are testable. *World Archaeology*, **34**, 26–46.

Bunn, H. T., Bartram, L. E. & Kroll, E. M. (1988). Variability in bone assemblage formation from Hadza hunting, scavenging, and carcass processing. *Journal of Anthropological Archaeology*, **7**, 412–457.

Bunn, H. T. & Ezzo, J. A. (1993). Hunting and scavenging by Plio-Pleistocene hominids: nutritional constraints, archaeological patterns, and behavioural implications. *Journal of Archaeological Science*, **20**, 365–398.

Burjachs, F. & Julia, R. (1994). Abrupt climatic changes during the last glaciation based on pollen analysis of the Abric Romani, Catalonia, Spain. *Quaternary Research*, **42**, 308–315.

Busack, S. D. (1986). Biogeographic analysis of the herpetofauna separated by the formation of the Strait of Gibraltar. *National Geographic Research*, **2**, 17–36.

Bush, A. B. G. & Philander, S. G. H. (1998). The Role of ocean-atmosphere interactions in tropical cooling during the Last Glacial Maximum. *Science*, **279**, 1341–1344.

Byers, A. M. (1994). Symboling and the Middle–Upper Palaeolithic transition: a theoretical and methodological critique. *Current Anthropology*, **35**, 369–399.

Bynum, N. (2002). Morphological variation within a macaque hybrid zone. *American Journal of Physical Anthropology*, **118**, 45–49.

Cabrera Valdes, V. & de Quiros, F. B. (1996). The origins of the Upper Palaeolithic: a Cantabrian perspective. In *The Last Neanderthals, The First Anatomically Modern Humans*, ed. E. Carbonell & M. Vaquero, pp. 251–265. Tarragona, Universitat Rovira i Virgili.

Cachel, S. (1997). Dietary shifts and the European Upper Palaeolithic transition. *Current Anthropology*, **38**, 579–603.

Cain, A. J. (1971). *Animal Species and their Evolution*. London, Hutchinson University Library.

Cann, R. L., Stoneking, M. L. & Wilson, A. C. (1987). Mitochondrial DNA and human evolution. *Nature*, **325**, 31–36.

Caramelli, D., Lalueza-Fox, D., Vernesi, C. *et al.* (2003). Evidence for a genetic discontinuity between Neandertals and 24,000-year-old anatomically modern Europeans. *Proceedings of the National Academy of Sciences, USA,* **100**: 6593–6597.

Carbonell, E., Bermúdez de Castro, J. M., Arsuaga, J. L. *et al.* (1995). Lower Pleistocene hominids and artefacts from Atapuerca-TD6 (Spain). *Science,* **269**, 826–832.

Carbonell, E., Mosquera, M., Rodriguez, X. P. & Sala, R. (1999a). Out of Africa: the dispersal of the earliest technical systems reconsidered. *Journal of Anthropological Archaeology,* **18**, 119–136.

Carbonell, E., García-Antón, M. D., Mallol, C. *et al.* (1999b). The TD6 level lithic industry from Gran Dolina, Atapuerca (Burgos, Spain): production and use. *Journal of Human Evolution,* **37**, 653–693.

Carcaillet, C. & Vernet, J-L. (2001). Comments on 'the full-glacial forests of central and southeastern Europe' by Willis *et al. Quaternary Research,* **55**, 385–387.

Carney, S. E., Gardner, K. A. & Riesberg, L. H. (2000). Evolutionary changes over the last fifty-year history of a hybrid population of sunflowers (*Helianthus*). *Evolution,* **54**, 462–474.

Carrier, D. (1984). The energetic paradox of human running and hominid evolution. *Current Anthropology,* **25**, 483–495.

Carrión, J. S. (1992). Late Quaternary pollen sequence from Carihuela Cave, southeastern Spain. *Review of Palaeobotany & Palynology,* **71**, 37–77.

Carrión, J. S., Dupré, M., Fumanal, M. P. & Montes, R. (1995). A Palaeoenvironmental study in semi-arid southeastern Spain: the palynological and sedimentological sequence at Perneras Cave (Lorca, Murcia). *Journal of Archaeological Science,* **22**, 355–367.

Carrión, J. S. & Munuera, M. (1997). Upper Pleistocene palaeoenvironmental change in Eastern Spain: new pollen-analytical data from Cova Beneito (Alicante). *Palaeogeography, Palaeoclimatology, Palaeoecology,* **128**, 287–299.

Carrión, J. S., Munuera, M. & Navarro, C. (1998). The palaeoenvironment of Carihuela Cave (Granada, Spain): a reconstruction on the basis of palynological investigations of cave sediments. *Review of Palaeobotany and Palynology,* **99**, 317–340.

Carrión, J. S., Munuera, M., Navarro, C. & Saez, F. (2000). Paleoclimas e historia de la vegetación cuaternaria en España a través del análisis polínico. Viejas falacias y nuevos paradigmas. *Complutum,* **11**, 115–142.

Cashdan, E. (1992). Spatial organisation and habitat use. In *Evolutionary Ecology and Human Behaviour,* ed. E. A. Smith & B. Winterhalder, pp. 237–266. New York, Aldine de Gruyter.

Cerdeño, E. (1998). Diversity and evolutionary trends of the Family Rhinocerotidae (Perissodactyla). *Palaeogeography, Palaeoclimatology, Palaeoecology,* **141**, 13–34.

Chase, P. G. & Dibble, H. (1987). Middle Palaeolithic symbolism: A review of current evidence and interpretations. *Journal of Anthropological Archaeology,* **6**, 263–293.

Chazan, M. (1995). The language hypothesis for the Middle-to-Upper Paleolithic transition: an examination based on a multiregional lithic analysis. *Current Anthropology,* **36**, 749–768.

Cheddadi, R., Mamakowa, K., Guiot, J., de Beaulieu, J-L., Reille, M., Andrieu, V., Granoszewski, W. & Peyron, O. (1998). Was the climate of the Eemian stable? A quantitative climate reconstruction from seven European pollen records. *Palaeogeography, Palaeoclimatology, Palaeoecology*, **143**, 73–85.

Chilardi, S., Frayer, D. W., Gioia, P., Macchiarelli, R. & Mussi, M. (1996). Fontana Nuova di Ragusa (Sicily, Italy): southernmost Aurignacian site in Europe. *Antiquity*, **70**, 553–563.

Churchill, S. E. (1998). Cold adaptation, heterochrony, and Neandertals. *Evolutionary Anthropology*, **7**, 46–60.

Cifelli, R. L. (1981). Patterns of evolution among the Artiodactyla and Perissodactyla (Mammalia). *Evolution*, **35**, 433–440.

Clark, P. U. & Mix, A. C. (2002). Ice sheets and sea level of the Last Glacial Maximum. *Quaternary Science Reviews*, **21**, 1–7.

CLIMAP Project Members (1976). The surface of the Ice-Age Earth. *Science*, **191**, 1131–1137.

Cody, M. L. (1974). *Competition and the Structure of Bird Communities*. Princeton, Princeton University Press.

Cody, M. L. (1975). Towards a theory of continental species diversities. In *Ecology and Evolution of Communities*, ed. M. L. Cody & J. M. Diamond, pp. 214–257. Cambridge, Massachusetts, Belknap Press.

Cody, M. L. (1986). Diversity, rarity, and conservation in Mediterranean-climate regions. In *Conservation Biology. The Science of Scarcity and Diversity*, ed. M. E. Soulé, pp. 123–152. Sunderland, Sinauer Associates.

COHMAP Members. (1988). Climate Changes of the Last 18,000 Years: Observations and Model Simulations. *Science*, **241**, 1043–1052.

Colinvaux, P. A., De Oliveira, P. E., Moreno, J. E., Miller, M. C. & Bush, M. B. (1996). A long pollen record from lowland Amazonia: forest and cooling in glacial times. *Science*, **274**, 85–88.

Cooke, H. B. S. (1968). The fossil mammal fauna of Africa. *Quarterly Review of Biology*, **43**, 234–264.

Cooke, H. B. S. (1978). Suid evolution and correlation of African hominid localities: an alternative taxonomy. *Science*, **201**, 460–463.

Coope, G. R. (2002). Changes in the thermal climate in northwestern Europe during marine Oxygen Isotope Stage 3, estimated from fossil insect assemblages. *Quaternary Research*, **57**, 401–408.

Courty, M-A. & Vallverdu, J. (2001). The microstratigraphic record of abrupt climate changes in cave sediments of the western Mediterranean. *Geoarchaeology*, **16**, 467–500.

Cox, C. B. & Moore, P. D. (1985). *Biogeography. An Ecological and Evolutionary Approach*. Oxford, Blackwell Scientific Publications.

Cox, G. W. & Ricklefs, R. E. (1977). Species diversity and ecological release in Caribbean land bird faunas. *Oikos*, **28**, 113–122.

Crombie, M. K., Arvidson, R. E., Sturchio, N. C., El Alfy, Z. & Abu Zeid, K. (1997). Age and isotopic constraints on Pleistocene pluvial episodes in the Western Desert, Egypt. *Palaeogeography, Palaeoclimatology, Palaeoecology*, **130**, 337–355.

Crow, T. J. (2002). Sexual selection, timing and an X-Y homologous gene: did *Homo sapiens* speciate on the Y chromosome? In *The Speciation of Modern Homo sapiens*, ed., T. J. Crow, pp. 197–216. Oxford, Oxford University Press.

Cuenca Bescós, G. (2003). The micromammal record as proxy of palaeoenvironmental changes in the Pleistocene of the Sierra de Atapuerca (Burgos, Spain). In *Quaternary climatic changes and environmental crises in the Mediterranean Region*, ed. M. B. Ruiz Zapata, M. Dorado Valiño, A. Valdeolmillos Rodríguez *et al.*, pp. 133–138. Alcalá de Henares, Universidad de Alcalá-MCYT-INQUA (Commission on Human Evolution and Palaeoecology).

Cuenca-Bescós, G., Laplana, C. & Canudo, J. I. (1999). Biochronological implications of the Arvicolidae (Rodentia, Mammalia) from the Lower Pleistocene hominid-bearing level of Trinchera Dolina 6 (TD6 Atapuerca, Spain). *Journal of Human Evolution*, **37**, 353–374.

Czarnetzki, A. (1995). Morphological evidence of adaptive characters in the genus *Homo*. In *Man and the environment in the Palaeolithic*, ed. H. Ullrich, pp. 97–110. Liege, Etudes et Recherches Archaeologiques de l'Université de Liege No. 62.

Dam, R. A. C., Fluin, J., Suparan, P. & van der Kaars, S. (2001). Palaeoenvironmental developments in the Lake Tondano area (N. Sulawesi, Indonesia) since 33,000 yr B.P. *Palaeogeography, Palaeoclimatology, Palaeoecology*, **171**, 147–183.

Danielson, B. J. (1991). Communities in a landscape: the influence of habitat heterogeneity on the interactions between species. *American Naturalist*, **138**, 1105–1120.

Danley, P. D., Markert, J. A., Arnegard, M. E. & Kocher, T. D. (2000). Divergence with gene flow in the rock-dwelling cichlids of Lake Malawi. *Evolution*, **54**, 1725–1737.

Dansgaard, W., Johnsen, S. J., Clausen, H. B. *et al.* (1993). Evidence for general instability of past climate from a 250-kyr ice-core record. *Nature*, **364**, 218–220.

Dansgaard, W., White, J. W. C. & Johnsen, S. J. (1989). The abrupt termination of the Younger Dryas climatic event. *Nature*, **339**, 532–534.

Dasmahapatra, K. K., Blum, M. J., Aiello, A. *et al.* (2002). Inferences from a rapidly moving hybrid zone. *Evolution*, **56**, 741–753.

Davis, R. S. & Ranov, V. A. (1999). Recent work on the Paleolithic of Central Asia. *Evolutionary Anthropology*, **8**, 186–193.

Davies, N. B. (1976). Food, flocking and territorial behaviour of the pied wagtail (*Motacilla alba yarellii* Gould) in winter. *Journal of Animal Ecology*, **45**, 235–253.

Dayan, T., Simberloff, D., Tchernov, E. & Yom-Tov, Y. (1991). Calibrating the paleothermometer: climate, communities, and the evolution of size. *Paleobiology*, **17**, 189–199.

Deacon, H. J. (1989). Late Pleistocene palaeoecology and archaeology in the southern Cape, South Africa. In *The Human Revolution*, ed. C. Stringer & P. Mellars, pp. 547–564. Edinburgh, Edinburgh University Press.

Dean, M. C., Stringer, C. B. & Bromage, T. C. (1986). Age at death of the Neanderthal child from Devil's Tower, Gibraltar and the implications for studies of general growth and development in Neanderthals. *American Journal of Physical Anthropology*, **70**, 301–310.

Debénath, A. (2000). Le peuplement préhistorique du Maroc: données récentes et problemes. *L'Anthropologie*, **104**, 131–145.

Demars, P-Y. (1996). Démographie et occupation de l'espace au Paléolithique supérieur et au Mésolithique en France. *Préhistoire Européenne*, **8**, 3–26.

Demars, P-Y. (1998). Circulation des silex dans le nord de l'Aquitaine au Paléolithique Supérieur. L'occupation de l'espace par les dernieres chasseurs-cueilleurs. *Gallia Préhistoire*, **40**, 1–28.

deMenocal, P. (1995). Plio-Pleistocene African Climate. *Science*, **270**, 53–59.

deMenocal, P. B. (2001). Cultural responses to climate change during the late Holocene. *Science*, **292**, 667–673.

Denton, G. H. (1999). Cenozoic climate change. In *African Biogeography, Climate Change, and Human Evolution*, ed. T. G. Bromage & F. Schrenk, pp. 94–114. Oxford, Oxford University Press.

Derbeneva, O. A., Starikovskaya, E. B., Wallace, D. C. & Sukernik, R. I. (2002). Traces of early Eurasians in the Mansi of Northwest Siberia revealed by mitochondrial DNA analysis. *American Journal of Human Genetics*, **70**, 1009–1014.

d'Errico, F., Henshilwood, C. & Nilssen, P. (2001). Engraved bone fragment from South Africa: implications for the origin of symbolism and language. *Antiquity*, **75**, 309–318.

d'Errico, F. & Nowell, A. (2000). A new look at the Berekhat ram figurine: implications for the origins of symbolism. *Cambridge Archaeological Journal*, **10**, 123–167.

d'Errico, F., Zilhao, J., Julien, M., Baffier, D. & Pelegrin, J. (1998). Neanderthal acculturation in western Europe? *Current Anthropology*, **39** (Suppl.), 1–44.

Diamond, J. (1991). *The Rise and Fall of the Third Chimpanzee*. London, Vintage.

Diamond, J. (1997). *Guns, Germs and Steel. The Fates of Human Societies*. London, Jonathan Cape.

Diamond, J. (2002). Evolution, consequences and future of plant and animal domestication. *Nature*, **418**, 700–707.

Dieckmann, U. & Doebeli, M. (1999). On the origin of species by sympatric speciation. *Nature*, **400**, 354–357.

Ditlevsen, P. D., Svensmark, H. & Johnsen, S. (1996). Contrasting atmospheric and climate dynamics of the last-glacial and Holocene periods. *Nature*, **379**, 810–812.

Duarte, C., Mauricio, J., Pettitt, P. B. *et al.* (1999). The early Upper Palaeolithic human skeleton from the Abrigo do Lagar Velho (Portugal) and modern human emergence in Iberia. *Proceedings of the National Academy of Sciences, USA*, **96**, 7604–7609.

du Boulay, G. H., Lawton, M. & Wallis, A. (1998). The story of the internal carotid artery of mammals: from Galen to sudden infant death syndrome. *Neuroradiology*, **40**, 697–703.

Dudley, J. P. (1996). Mammoths, gomphotheres, and the Great American Faunal Interchange. In *The Proboscidea. Evolution and Palaeoecology of Elephants and Their Relatives*, ed. J. Shoshani & P. Tassy, pp. 289–295. Oxford, Oxford University Press.

Dunbar, R. (1984). *Reproductive decisions*. Princeton, Princeton University Press.

Dunbar, R. I. M. (1992). Neocortex size as a constraint on group size in primates. *Journal of Human Evolution*, **20**, 469–493.

Dunnell, R. C. (1999). The concept of waste in an evolutionary archaeology. *Journal of Anthropological Archaeology*, **18**, 243–250.

Dunnell, R. C. & Greenlee, D. M. (1999). Late Woodland Period 'waste' reduction in the Ohio River Valley. *Journal of Anthropological Archaeology*, **18**, 376–395.

Dupont, L. M. (1993). Vegetation zones in NW Africa during the Brunhes Chron reconstructed from marine palynological data. *Quaternary Science Reviews*, **12**, 189–202.

Dupont, L., Jahns, S., Marret, F. & Ning, S. (2000). Vegetation change in equatorial West Africa: time-slices for the last 150 ka. *Palaeogeography, Palaeoclimatology, Palaeoecology*, **155**, 95–122.

Dynesius, M. & Jansson, R. (2000). Evolutionary consequences of changes in species' geographical distributions driven by Milankovitch climate oscillations. *Proceedings of the National Academy of Sciences, USA*, **97**, 9115–9120.

Dytham, C. (2000). Habitat destruction and extinctions: predictions from metapopulation models. In *The Ecological Consequences of Environmental Heterogeneity*, ed. M. J. Hutchings, E. A. John & A. J. A. Stewart, pp. 315–332. Oxford, Blackwell Science.

Eisenberg, J. F. (1981). *The Mammalian Radiations. An Analysis of Trends in Evolution, Adaptation, and Behaviour*. London, Athlone Press.

Elliot, M., Labeyrie, L. & Duplessy, J.-C. (2002). Changes in North Atlantic deepwater formation associated with the Dansgaard–Oeschger temperature oscillations (60–10 ka). *Quaternary Science Reviews*, **21**, 1153–1165.

Elson, J. L., Andrews, R. M., Chinnery, P. F., Lightowlers, R. N., Turnbull, D. M. & Howell, N. (2001). Analysis of European mtDNAs for recombination. *American Journal of Human Genetics*, **68**, 145–153.

Elston, R. G. & Zeanah, D. W. (2002). Thinking outside the box: a new perspective on diet breadth and sexual division of labor in the Great Basin. *World Archaeology*, **34**, 103–130.

Elton, S., Bishop, L. C. & Wood, B. (2001). Comparative context of Plio-Pleistocene hominin brain evolution. *Journal of Human Evolution*, **41**, 1–27.

Enard, W., Przeworski, M., Fisher, S. E. *et al.* (2002). Molecular evolution of FOXP2, a gene involved in speech and language. *Nature*, **418**, 869–872.

Endler, J. A. (1992). Signals, signal conditions, and the direction of evolution. *American Naturalist*, **74**, 312–321.

Excoffier, L. & Schneider, S. (1999). Why hunter–gatherer populations do not show signs of Pleistocene demographic expansions. *Proceedings of the National Academy of Sciences, USA*, **96**, 10597–10602.

Eyre-Walker, A., Smith, N. H. & Maynard Smith, J. (1999). How clonal are human mitochondria? *Proceedings of the Royal Society, Series B*, **266**, 477–484.

Fa, D., Finlayson, C., Gils Pacheco, F., Finlayson, G., Aguilera Garcia, J. & Aguilera Garcia, J. A. (2001). Building bridges: new perspectives on Out-of-Africa. In *Neanderthals and Modern Humans in Late Pleistocene Eurasia*, ed. C. Finlayson, pp. 31–34. Gibraltar: Gibraltar Government Heritage Publications.

Falgueres, C., Bahain, J-J, Yokoyama, Y. *et al.* (1999). Earliest humans in Europe: the age of TD6 Gran Dolina, Atapuerca, Spain. *Journal of Human Evolution*, **37**, 343–352.

FAUNMAP Working Group (1996). Spatial response of mammals to late Quaternary environmental fluctuations. *Science*, **272**, 1601–1606.

Féblot-Augustins, J. (1990). Exploitation des matieres premieres dans l'Acheuléen d'Afrique: perspectives comportementales. *Paléo*, **2**, 27–42.

Féblot-Augustins, J. (1993). Mobility strategies in the late Middle Palaeolithic of Central Europe and Western Europe: elements of stability and variability. *Journal of Anthropological Archaeology*, **12**, 211–265.

Fehr, E. & Gachter, S. (2002). Altruistic punishment in humans. *Nature*, **415**, 137–140.

Fernández-Armesto, F. (2000). *Civilizations*. London, Macmillan.

Ferguson, S. H., Taylor, M. K., Born, E. W. & Messier, F. (1998). Fractals, sea-ice landscape and spatial patterns of polar bears. *Journal of Biogeography*, **25**, 1081–1092.

Ferguson, S. H., Taylor, M. K., Born, E. W., Rosing-Asvid, A. & Messier, F. (1999). Determinants of home range size for polar bears (*Ursus maritimus*). *Ecology Letters*, **2**, 311–318.

Field, C. B., Behrenfeld, M. J., Randerson, J. T. & Falkowski, P. (1998). Primary production of the biosphere: integrating terrestrial and oceanic components. *Science*, **281**, 237–240.

Figueiral, I. & Terral, J.-F. (2002). Late Quaternary refugia of Mediterranean taxa in the Portuguese Estremadura: charcoal based palaeovegetation and climatic reconstruction. *Quaternary Science Reviews*, **21**, 549–558.

Filchak, K. E., Roethele, J. B. & Feder, J. L. (2000). Natural selection and sympatric divergence in the apple maggot *Rhagoletis pomonella*. *Nature*, **407**, 739–742.

Finlayson, C. (1999). Late Pleistocene human occupation of the Iberian Peninsula. *Journal of Iberian Archaeology*, **1**, 59–68.

Finlayson, C. (2003). The role of climate in the spatio-temporal pattern of human colonization and extinction in the Pleistocene with specific reference to the Mediterranean Region. In *Quaternary Climatic Changes and Environmental Crises in the Mediterranean Region*, ed. M. B. Ruiz Zapata, M. Dorado Valiño, A. Valdeolmillos Rodríguez *et al.*, pp. 57–65. Alcalá de Henares, Universidad de Alcalá-MCYT-INQUA (Commission on Human Evolution and Palaeoecology).

Finlayson, C., Fa, D. A. & Finlayson, G. (2000a). Biogeography of human colonizations and extinctions in the Pleistocene. *Memoirs Gibcemed*, **1**, 1–69.

Finlayson, C. & Giles Pacheco, F. (2000). The southern Iberian Peninsula in the late Pleistocene: geography, ecology and human occupation. In *Neanderthals on the Edge*, ed. C. B. Stringer, R. N. E. Barton, & J. C. Finlayson, pp. 139–154. Oxford, Oxbow Books.

Finlayson, C., Giles Pacheco, F., Gutierrez López, J. M. *et al.* (1999). Recientes excavaciones en el nivel neolítico de la cueva de Gorham (Gibraltar. Extremo Sur de Europa). *Saguntum*, Extra-**2**, 213–222.

Finlayson, J. C., Barton, R. N. E., Giles Pacheco, F. *et al.* (2000b). Human occupation of Gibraltar during Oxygen Isotope Stages 2 and 3 and a comment on the late survival of Neanderthals in the southern Iberian Peninsula. *Paleolítico da Península Ibérica*, **2**, 277–285.

Fisher, D. C. (1996). Extinction of proboscideans in North America. In *The Proboscidea. Evolution and Palaeoecology of Elephants and Their Relatives*, ed. J. Shoshani & P. Tassy, pp. 296–315. Oxford, Oxford University Press.

Fitzhugh, B. (2001). Risk and invention in human technological evolution. *Journal of Anthropological Archaeology*, **20**, 125–167.

Fizet, M., Mariotti, A., Bocherens, H., Lange-Badré, B. *et al.* (1995). Effect of diet, physiology and climate on carbon and nitrogen stable isotopes of collagen in a late Pleistocene anthropic palaecosystem: Marillac, Charente, France. *Journal of Archaeological Science*, **22**, 67–79.

Flohn, H. (1984). A possible mechanism of abrupt climatic changes. In *Climatic Changes on a Yearly to Millennial Basis*, ed. N. A. Morner & W. Karlen, pp. 521–532. Dordrecht, Reidel.

Florschutz, F., Menendez Amor, J. & Wijmstra, T. A. (1971). Palynology of a thick Quaternary succession in southern Spain. *Palaeogeography, Palaeoclimatology, Palaeoecology*, **10**, 233–264.

Flynn, L. J., Tedford, R. H. & Qiu, Z. (1991). Enrichment and stability in the Pliocene mammalian fauna of North China. *Paleobiology*, **17**, 246–265.

Foley, P. (1997). Extinction models for local populations. In *Metapopulation Biology. Ecology, Genetics, and Evolution*, ed. I. A. Hanski & M. E. Gilpin, pp. 215–246. San Diego, Academic Press.

Foley, R. (1987). *Another Unique Species. Patterns in Human Evolutionary Ecology.* Harlow, Longman.

Foley, R. (1989). The ecological conditions of speciation: a comparative approach to the origins of anatomically modern humans. In *The Human Revolution*, ed. P. Mellars & C. B. Stringer, pp. 298–318. Edinburgh, Edinburgh University Press.

Foley, R. (1992). Evolutionary ecology of fossil hominids. In *Evolutionary Ecology and Human Behaviour*, ed. E. A. Smith & B. Winterhalder, pp. 131–164. New York, Aldine de Gruyter.

Foley, R. (1994). Speciation, extinction and climatic change in hominid evolution. *Journal of Human Evolution*, **26**, 275–289.

Foley, R. (1998). Genes, evolution, and diversity: yet another look at the problem of modern human origins. *Evolutionary Anthropology*, **6**, 191–193.

Foley, R. (2001). In the shadow of the modern synthesis? Alternative perspectives on the last fifty years of paleoanthropology. *Evolutionary Anthropology*, **10**, 5–14.

Foley, R. & Lahr, M. M. (1997). Mode 3 technologies and the evolution of modern humans. *Cambridge Archaeology*, **7**, 3–36.

Foley, R. A. & Lee, P. C. (1989). Finite social space, evolutionary pathways, and reconstructing hominid behaviour. *Science*, **243**, 901–906.

Forman, R. T. T. (1995). *Land Mosaics. The Ecology of Landscapes and Regions.* Cambridge, Cambridge University Press.

Fortea, J. (1973). *Los complejos microlaminares y geométricos del Epipaleolítico Mediterráneo español*. Memorias del Seminario de Arqueología. Salamanca, Universidad de Salamanca.

Fortea Pérez, F. J. (1978). Arte paleolítico del Mediterráneo español. *Trabajos de Prehistoria*, **35**, 99–149.

Franciscus, R. G. & Trinkaus, E. (1988). Nasal morphology and the emergence of *Homo erectus. American Journal of Physical Anthropology*, **75**, 517–527.

Frayer, D. W. (1984). Biological and cultural change in the European late Pleistocene and early Holocene. In *The Origin of Modern Humans: A World Survey of the Fossil Evidence*, ed. F. H. Smith & F. Spencer, pp. 211–250. New York, Alan R. Liss.

Frison, G. C. (1989). Experimental use of Clovis weaponry and tools on African elephants. *American Antiquity*, **54**, 766–784.

Frison, G. C. (1998). Paleoindian large mammal hunters on the plains of North America. *Proceedings of the National Academy of Sciences, USA*, **95**, 14576–14583.

Fritz, H. & Duncan, P. (1994). On the carrying capacity for large ungulates of African savanna ecosystems. *Proceedings of the Royal Society London, Series B*, **256**, 77–82.

Frogley, M. R., Tzedakis, P. C. & Heaton, T. H. E. (1999). Climate Variability in northwest Greece during the last interglacial. *Science*, **285**, 1886–1889.

Fryxell, J. M. (1995). Aggregation and migration by grazing ungulates in relation to resources and predators. In *Serengeti II. Dynamics, Management, and Conservation of an Ecosystem*, ed. A. R. E. Sinclair & P. Arcese, pp. 257–273. Chicago, University of Chicago Press.

Gabunia, L., Vekua, A., Lordkipanidze, D. *et al.* (2000). Earliest Pleistocene hominid cranial remains from Dmanisi, Republic of Georgia: taxonomy, geological setting, and age. *Science*, **288**, 1019–1025.

Gabunia, L., Antón, S. C., Lordkipanidze, D., Vekua, A., Justus, A. & Swisher III, C. C. (2001). Dmanisi and dispersal. *Evolutionary Anthropology*, **10**, 158–170.

Gagneux, P. & Varki, A. (2001). Genetic differences between humans and great apes. *Molecular Phylogenetics and Evolution*, **18**, 2–13.

Gamble, C. (1986). *The Palaeolithic Settlement of Europe*. Cambridge, Cambridge University Press.

Gamble, C. (1993). *Timewalkers. The Prehistory of Global Colonization*. London, Penguin.

Gamble, C. (1995). Large mammals, climate and resource richness in Upper Pleistocene Europe. *Acta. Zoologica Cracovia*, **38**, 155–175.

Gamble, C. (1999). *The Palaeolithic Societies of Europe*. Cambridge, Cambridge University Press.

Gamble, C. (2001). Modes, movement and moderns. *Quaternary International*, **75**, 5–10.

Garcia, N. & Arsuaga, J. L. (2003). Late Pleistocene cold-resistant faunal complex: Iberian occurrences. *In Quaternary Climatic Changes and Environmental Crises in the Mediterranean Region*, ed. M. Blanca Ruiz Zapata, M. Dorado Valino, A. Valdeolmillos Rodriguez *et al.*, pp. 149–159. Madrid, Universidad de Alcala de Henares.

García-Ramos, G. & Rodríguez, D. (2002). Evolutionary speed of species invasions. *Evolution*, **56**, 661–668.

Gargett, R. H. (1989). Grave shortcomings: the evidence for Neanderthal burial. *Current Anthropology*, **30**, 157–190.

Gasse, F. (2000). Hydrological changes in the African tropics since the Last Glacial Maximum. *Quaternary Science Reviews*, **19**, 189–211.

Gaven, C., Hillaire-Marcel, C. & Petit-Maire, N. (1981). A Pleistocene lacustrine episode in southeastern Libya. *Nature*, **290**, 131–133.

Gavrilets, S., Li, H. & Vose, M. D. (1998). Rapid parapatric speciation on holey adaptive landscapes. *Proceedings of the Royal Society London, Series B*, **265**, 1483–1489.

Geist, V. (1977). A comparison of social adaptations in relation to ecology in gallinaceous bird and ungulate societies. *Annual Review of Ecology and Systematics*, **8**, 193–207.

Geist, V. (1998). *Deer of the World. Their Evolution, Behavior, and Ecology*. Mechanicsburg, Stackpole Books.

Giles, F., Gutiérrez, J. Ma., Santiago, A., Mata, E. & Gracia, F. J. (1993). Prospecciones arqueológicas y análisis geocronólogicos y sedimentológicos en la cuenca del río Guadalete. Secuencia fluvial y paleolítica del río Guadalete (Cádiz). Resultados de las investigaciones hasta 1993. In *Investigaciones Arqueológicas en Andalucía 1985–1992*, Proyectos, 211–227. Sevilla.

Giles Pacheco, F. & Santiago Pérez, A. (1987). El Poblamiento del sur de la Peninsula Ibérica en el Pleistoceno Inferior a través de Gibraltar. In *Actas del Congreso Internacional 'El Estrecho de Gibraltar'* ed. E. Ripoll Perelló, vol. 1, pp. 97–109. Ceuta, UNED.

Giles Pacheco, F., Finlayson, C., Santiago Pérez, A., Fa, D., Gutiérrez López, J. M. & Finlayson, G. (2003). The effect of climate change on the distribution of humans in southern Iberia in the late Quaternary. In *Quaternary Climatic Changes and Environmental Crises in the Mediterranean Region*, ed. M. B. Ruiz Zapata, M. Dorado Valiño, A. Valdeolmillos Rodríguez *et al.*, pp. 67–79. Alcalá de Henares, Universidad de Alcalá-MCYT-INQUA (Commission on Human Evolution and Palaeoecology).

Gillman, H. & Hails, R. (1997). *An Introduction to Ecological Modelling. Putting Practice into Theory*. Oxford, Blackwell.

Gilpin, M. E. & Soulé, M. E. (1986). Minimum viable populations: processes of species extinction. In *Conservation Biology. The Science of Scarcity and Diversity*, ed. M. E. Soulé, pp. 19–34. Sunderland, Sinauer Associates.

Gisolfi, C. V. & Mora, F. (2000). *The Hot Brain. Survival, Temperature, and the Human Body*. Cambridge, Massachussetts, MIT Press.

Goebel, T. (1999). Pleistocene human colonization of Siberia and peopling of the Americas: an ecological approach. *Evolutionary Anthropology*, **8**, 208–227.

Goebel, T., Waters, M. R., Buvit, I., Konstantinov, M. V. & Konstantinov, A. V. (2000). Studenoe-2 and the origins of microblade technologies in the Transbaikal, Siberia. *Antiquity*, **74**, 567–575.

Goodfriend, G. A. (1999). Terrestrial stable isotope records of late Quaternary paleoclimates in the eastern Mediterranean region. *Quaternary Science Reviews*, **18**, 501–513.

Gould, S. J. & Eldredge, N. (1977). Punctuated equilibria: the tempo and mode of evolution reconsidered. *Paleobiology*, **3**, 115–151.

Grayson, D. K. & Delpech, F. (2002). Specialized early Upper Palaeolithic hunters in southwestern France? *Journal of Archaeological Science*, **29**, 1439–1449.

GRIP Members (1993). Climate instability during the last interglacial period recorded in the GRIP ice core. *Nature*, **364**, 203–207.

Groom, M. J. & Pascual, M. A. (1998). The analysis of population persistence: an outlook on the practice of viability analysis. In *Conservation Biology for the Coming Decade*, ed. P. L. Fiedler & P. M. Karieva, pp. 4–27. New York, Chapman & Hall.

Grove, A. T. & Warren, A. (1968). Quaternary landforms and climate on the south side of the Sahara. *Geographical Journal*, **134**, 194–208.

Guérin, C. & Patou-Mathis, M. (1996). *Les Grands Mammiferes Plio-Pleistocenes d'Europe.* Paris, Masson.

Guiot, J., Pons, A., de Beaulieu, J. L. & Reille, M. (1989). A 140,000-year continental climate reconstruction from two European pollen records. *Nature,* **338,** 309–313.

Guthrie, R. D. (1984). Mosaics, allelochemics, and nutrients: an ecological theory of late Pleistocene megafaunal extinctions. In *Quaternary Extinctions. A Prehistoric Revolution,* ed. P. S. Martin & R. G. Klein, pp. 259–298. Tucson, University of Arizona Press.

Gvirtzman, G. & Wieder, M. (2001). Climate of the last 53,000 years in the eastern Mediterranean, based on soil-sequence stratigraphy in the coastal plain of Israel. *Quaternary Science Reviews,* **20,** 1827–1849.

Gyllenberg, M., Hanski, I. & Hastings, A. (1997). Structured metapopulation models. In *Metapopulation Biology. Ecology, Genetics, and Evolution,* ed. I. A. Hanski & M. E. Gilpin, pp. 93–122. San Diego, Academic Press.

Hagelberg, E., Goldman, N., Lió, P., Whelan, S., Schiefenhovel, W., Clegg, J. B. & Bowden, D. K. (1999). Evidence for mitochondrial DNA recombination in a human population of island Melanesia. *Proceedings of the Royal Society of London, Series B,* **266,** 485–492.

Haigh, J. & Maynard Smith, J. (1972). Population size and protein variation in man. *Genetic Research,* **19,** 73–89.

Hamilton, F. E. (1999). Southeastern archaic mounds: examples of elaboration in a temporally fluctuating environment? *Journal of Anthropological Archaeology,* **18,** 344–355.

Hamilton, W. D. (1964). The genetical theory of social behaviour. I, II. *Journal of Theoretical Biology,* **7,** 1–52.

Hamilton, W. R. (1978). Fossil giraffes from the Miocene of Africa and a revision of the phylogeny of the Giraffoidea. *Philosophical Transactions of the Royal Society London, Series B,* **283,** 165–229.

Hammer, M. F., Karafet, T., Rasanayagam, A. *et al.* (1998). Out of Africa and back again: nested cladistic analysis of human Y chromosome variation. *Molecular Biology and Evolution,* **15,** 427–441.

Hammer, M. F., Spurdle, A. B., Karafet, T. *et al.* (1997). The geographic distribution of human Y chromosome variation. *Genetics,* **145,** 787–805.

Hanski, I. (1983). Coexistence and competitors in a patchy environment. *Ecology,* **64,** 493–500.

Hanski, I. A. & Gilpin, M. E. (Ed.) (1997). *Metapopulation Biology. Ecology, Genetics, and Evolution.* San Diego, Academic Press.

Harding, R. M., Fullerton, S. M., Griffiths, R. C. & Clegg, J. B. (1997). A gene tree for beta-globin sequences from Melanesia. *Journal of Molecular Evolution,* **44,** S133–S138.

Harpending, H. C., Batzer, M. A., Gurven, M., Jorde, L. B., Rogers, A. R. & Sherry, S. T. (1998). Genetic traces of ancient demography. *Proceedings of the National Academy of Sciences, USA,* **95,** 1961–1967.

Harpending, H. C. & Davis, H. (1977). Some implications for hunter–gatherer ecology derived from the spatial structure of resources. *World Archaeology,* **8,** 275–286.

Harpending, H. & Rogers, A. (2000). Genetic perspectives on human origins and differentiation. *Annual Review of Genomics and Human Genetics*, **1**, 361–385.

Harpending, H., Sherry, S. T., Rogers, A. L. & Stoneking, M. (1993). The genetic structure of ancient human populations. *Current Anthropology*, **34**, 483–496.

Harris, E. E. & Hey, J. (1999). X Chromosome evidence for ancient human histories. *Proceedings of the National Academy of Sciences, USA*, **96**, 3320–3324.

Harrison, S. (1991). Local extinction in a metapopulation context: an empirical evaluation. *Biological Journal of the Linnean Society*, **42**, 73–88.

Haskell, J. P., Ritchie, M. E. & Olff, H. (2002). Fractal geometry predicts varying body size scaling relationships for mammal and bird home ranges. *Nature*, **418**, 527–530.

Hawks, J., Hunley, K., Lee, S-H. & Wolpoff, M. (2000). Population bottlenecks and Pleistocene human evolution. *Molecular Biology and Evolution*, **17**, 2–22.

Hawks, J. D. & Wolpoff, M. H. (2001). The accretion model of Neandertal evolution. *Evolution*, **55**, 1474–1485.

Hayden, B. (1993). The cultural capacities of Neandertals: a review and re-evaluation. *Journal of Human Evolution*, **24**, 113–146.

Haynes, G. (1991). *Mammoths, mastodonts, and elephants: biology, behavior, and the fossil record*. Cambridge, Cambridge University Press.

Heinrich, H. (1988). Origin and consequences of cyclic ice rafting in the northeast Atlantic during the past 130,000 years. *Quaternary Research*, **28**, 142–152.

Helmke, J. P., Schulz, M. & Bauch, H. A. (2002). Sediment-color record from the Northeast Atlantic reveals patterns of millennial-scale climate variability during the past 500,000 years. *Quaternary Research*, **57**, 49–57.

Henrich, J. & Boyd, R. (2001). Why people punish defectors. *Journal of Theoretical Biology*, **208**, 79–89.

Henshilwood, C. S., d'Errico, F., Yates, R. *et al.* (2002). Emergence of modern human behaviour: Middle Stone Age engravings from South Africa. *Science*, **295**, 1278–1280.

Henshilwood, C. & Sealy, J. (1997). Bone artefacts from the Middle Stone Age at Blombos Cave, Southern Cape, South Africa. *Current Anthropology*, **38**, 890–895.

Hewitt, G. M. (1989). The subdivision of species by hybrid zones. In *Speciation and its Consequences*, ed. D. Otte & J. A. Endler, pp. 85–110. Sunderland, Sinauer.

Hewitt, G. M. (1996). Some genetic consequences of ice ages and their role in divergence and speciation. *Biological Journal of the Linnean Society*, **58**, 247–276.

Hewitt, G. M. (1999). Post-glacial re-colonization of European biota. *Biological Journal of the Linnean Society*, **68**, 87–112.

Hewitt, G. M. (2000). The genetic legacy of the Quaternary ice ages. *Nature*, **405**, 907–913.

Hill, R. A. & Dunbar, R. I. M. (2002). Climatic determinants of diet and foraging behaviour in baboons. *Evolutionary Ecology*, **16**, 579–593.

Hitchcock, R. K. (1989). Settlement, seasonality, and subsistence stress among the Tyva of northern Botswana. In *Coping with Seasonal Constraints*, ed. R. Huss-Ashmore, J. J. Curry & R. K. Hitchcock, pp. 64–85. Philadelphia, University of Pennsylvania.

Holliday, T. W. (1997a). Postcranial evidence of cold adaptation in European Nean-derthals. *American Journal of Physical Anthropology*, **104**, 245–258.

Holliday, T. W. (1997b). Body proportions in Late Pleistocene Europe and modern human origins. *Journal of Human Evolution*, **32**, 423–447.

Holliday, T. W. (1998). The ecological context of trapping among recent hunter–gatherers: implications for subsistence in terminal Pleistocene Europe. *Current Anthropology*, **39**, 711–719.

Holliday, T. W. & Falsetti, A. B. (1995). Lower limb length of European early modern humans in relation to mobility and climate. *Journal of Human Evolution*, **29**, 141–153.

Hooghiemstra, H., Stalling, H., Agwu, C. O. C. & Dupont, L. M. (1992). Vegetational and climatic changes at the northern fringe of the Sahara 250,000–5000 years BP: evidence from 4 marine pollen records located between Portugal and the Canary Islands. *Review of Palaeobotany and Palynology*, **74**, 1–53.

Hou, Y., Potts, R., Yuan, B. *et al.* (2000). Mid-Pleistocene Acheulian-like stone tech-nology of the Bose Basin, South China. *Science*, **287**, 1622–1626.

Housley, R. A., Gamble, C. S., Street, M. & Pettitt, P. (1997). Radiocarbon evidence for the late glacial human recolonisation of northern Europe. *Proceedings of the Prehistoric Society*, **63**, 25–54.

Hublin, J. J. (1992). Recent human evolution in Northwestern Africa. *Philosophical Transactions of the Royal Society, Series B*, **337**, 185–191.

Hublin, J-J. (1998). Climatic changes, paleogeography, and the evolution of the Ne-andertals. In *Neandertals and Modern Humans in Western Asia*, ed. T. Akazawa, K. Aochi & O. Bar-Yosef, pp. 295–310. New York, Plenum Press.

Hughen, K. A., Overpeck, J. T., Peterson, L. C. & Trumbore, S. (1996). Rapid climate changes in the tropical Atlantic region during the last deglaciation. *Nature*, **380**, 51–54.

Hunt, K. D. (1994). The evolution of human bipedality: ecology and functional mor-phology. *Journal of Human Evolution*, **26**, 183–202.

Huntley, B. & Birks, H. J. B. (1983). *An atlas of past and present pollen maps of Europe*. Cambridge, Cambridge University Press.

Hutchings, M. J., Wijesinghe, D. K. & John, E. A. (2000). The effects of heterogeneous nutrient supply on plant performance: a survey of responses, with special reference to clonal herbs. In *The Ecological Consequences of Environmental Heterogeneity*, ed. M. J. Hutchings, E. A. John & A. J. A. Stewart, pp. 91–110. Oxford, Blackwell Science.

Hutchinson, G. E. (1959). Homage to Santa Rosalia, or why are there so many kinds of animals? *American Naturalist*, **93**, 145–159.

Imbrie, J., Boyle, E. A., Clemens, S. C. *et al.* (1992). On the structure and origin of major glaciation cycles. *Palaeoceanography*, **7**, 701–738.

Imbrie, J., Hays, J. D., Martinson, A. *et al.* (1984). The orbital theory of Pleistocene climate: support from a revised chronology of the $\delta^{18}O$ record. In *Milankovitch and Climate*, ed. A. Berger, J. Imbrie, J. Hays, G. Kukla & B. Saltzman, pp. 269–306. Dordrecht, Reidel.

Ingman, M., Kaessmann, H., Paabo, S. & Gyllensten, U. (2000). Mitochondrial genome variation and the origin of modern humans. *Nature*, **408**, 708–713.

Irwin, D. E. (2000). Song variation in an avian ring species. *Evolution*, **54**, 998–1010.

Irwin, D. E., Irwin, J. H. & Price, T. D. (2001). Ring species as bridges between microevolution and speciation. *Genetica*, **112–113**, 223–243.

Isbell, L. A. & Young, T. P. (1996). The evolution of bipedalism in hominids and reduced group size in chimpanzees: alternative responses to decreasing resource availability. *Journal of Human Evolution*, **30**, 389–397.

Jablonski, N. G. & Chaplin, G. (1993). Origin of habitual terrestrial bipedalism in the ancestor of the Hominidae. *Journal of Human Evolution*, **24**, 259–280.

Jackson, A. P., Eastwood, H., Bell, S. M. *et al.* (2002). Identification of Microcephalin, a protein implicated in determining the size of the human brain. *American Journal of Human Genetics*, **71**, 136–142.

Jahns, S., Huls, M. & Sarnthein, M. (1998). Vegetation and climate history of west equatorial Africa based on a marine pollen record off Liberia (site GIK 16776) covering the last 400,000 years. *Review of Palaeobotany and Palynology*, **102**, 277–288.

Jarman, P. J. & Sinclair, A. R. E. (1979). Feeding strategy and the pattern of resource-partitioning in ungulates. In *Serengeti. Dynamics of an Ecosystem*, ed. A. R. E. Sinclair & M. Norton-Griffiths, pp. 130–163. Chicago, University of Chicago Press.

Jin, L., Underhill, P. A., Doctor, V. *et al.* (1999). Distribution of haplotypes from a chromosome 21 region distinguishes multiple prehistoric human migrations. *Proceedings of the National Academy of Sciences, USA*, **96**, 3796–3800.

Johannesson, K. (2001). Parallel speciation: a key to sympatric divergence. *Trends in Ecology and Evolution*, **16**, 148–153.

Johnson, T. C., Brown, E. T., McManus, J., Barry, S., Barker, P. & Gasse, F. (2002). A high-resolution paleoclimate record spanning the past 25,000 years in southern East Africa. *Science*, **296**, 113–132.

Jones, J. S. & Rouhani, S. (1986). How small was the bottleneck? *Nature*, **319**, 449–450.

Jorde, L.B., Bamshad, M. & Rogers, A. R. (1998). Using mitochondrial and nuclear DNA markers to reconstruct human evolution. *Bioessays*, **20**, 126–136.

Jorde, L. B., Rogers, A. R., Bamshad, M. *et al.* (1997). Microsatellite diversity and the demographic history of modern humans. *Proceedings of the National Academy of Sciences, USA*, **94**, 3100–3103.

Jorde, L. B., Watkins, W. S., Bamshad, M. J. *et al.* (2000). The distribution of human genetic diversity: a comparison of mitochondrial, autosomal, and Y-chromosome data. *American Journal of Human Genetics*, **66**, 979–988.

Kalb, J. E., Froehlich, D. J. & Bell, G. L. (1996). Palaeobiogeography of late Neogene African and Eurasian Elephantoidea. In *The Proboscidea. Evolution and Palaeoecology of Elephants and Their Relatives*, ed. J. Shoshani & P. Tassy, pp. 117–123. Oxford, Oxford University Press.

Kaessmann, H., Heibig, F., von Haesler, A. & Paabo, S. (1999). DNA sequence variation in a non-coding region of low recombination on the human X chromosome. *Nature Genetics*, **22**, 78–81.

Kaplan, H. & Hill, K. (1985). Food-sharing among Ache foragers: tests of explanatory hypotheses. *Current Anthropology*, **26**, 223–245.

Kaplan, H., Hill, K., Lancaster, J. & Hurtado, A. M. (2000). A theory of human life history evolution: diet, intelligence, and longevity. *Evolutionary Anthropology*, **9**, 156–185.

Kaplan, H. & Hill, K. (1992). The evolutionary ecology of food acquisition. In *Evolutionary Ecology and Human Behaviour*, ed. E. A. Smith & B. Winterhalder, pp. 167–201. New York, Aldine de Gruyter.

Kappelman, J. (1996). The evolution of body mass and relative brain size in fossil hominids. *Journal of Human Evolution*, **30**, 243–276.

Karavanich, I. & Smith, F. H. (1998). The Middle/Upper Palaeolithic interface and the relationship of Neanderthals and early modern humans in the Hrvatsko Zagorje, Croatia. *Journal of Human Evolution*, **34**, 223–248.

Karieva, P. (1990). Population dynamics in spatially complex environments: theory and data. *Philosophical Transactions of the Royal Society, London, Series B*, **330**, 175–190.

Kelly, R. L. (1983). Hunter–gatherer mobility strategies. *Journal of Anthropological Research*, **39**, 277–306.

Kelly, R. L. (1992). Mobility/sedentism: concepts, archaeological measures, and effects. *Annual Review of Anthropology*, **21**, 43–66.

Kelly, R. L. & Todd, L. C. (1988). Coming into the country: early paleoindian mobility and hunting. *American Antiquity*, **53**, 231–244.

Kent, M. & Coker, P. (1992). *Vegetation Description and Analysis. A Practical Approach.* Chichester, John Wiley & Sons.

Kerr, J. T. & Packer, L. (1997). Habitat heterogeneity as a determinant of mammal species richness in high-energy regions. *Nature*, **385**, 252–254.

Kingston, J. D., Marino, B. D. & Hill, A. (1994). Isotopic evidence for Neogene hominid paleoenvironments in the Kenya Rift Valley. *Science* **264**, 955–959.

Kirkpatrick, M. & Barton, N. H. (1997). Evolution of a species' range. *The American Naturalist*, **150**, 1–23.

Klein, R. (1995). Anatomy, behaviour, and modern human origins. *Journal of World Prehistory*, **9**, 167–198.

Klein, R. (1999). *The Human Career. Human Biological and Cultural Origins.* Chicago, University of Chicago Press.

Klein, R. G. (2000). Archeology and the evolution of human behavior. *Evolutionary Anthropology*, **9**, 17–36.

Klicka, J. & Zink, R. M. (1997). The importance of recent Ice Ages in speciation: a failed paradigm. *Science*, **277**, 1666–1669.

Kondrashov, A. S. & Kondrashov, F. A. (1999). Interactions among quantitative traits in the course of sympatric speciation. *Nature*, **400**, 351–354.

Kornbacher, K. D. (1999). Cultural elaboration in prehistoric coastal Peru: an example of evolution in a temporally variable environment. *Journal of Anthropological Archaeology*, **18**, 282–318.

Koumouzelis, M., Kozlowski, J. K., Escutenaire, C. *et al.* (2001). Le fin du Paléolithique moyen et le début du Paléolithique supérieur en Grece: la sequence de la Grotte 1 de Klissoura. *L'Anthropologie*, **105**, 469–504.

Kratzing, C. C. & Cross, R. B. (1984). Effects of facial cooling during exercise at high temperature. *European Journal of Applied Physiology*, **53**, 118–120.

Krings, M., Capelli, C., Tschentscher, F. *et al.* (2000). A view of Neandertal genetic diversity. *Nature Genetics*, **26**, 144–146.

Krings, M., Geisert, H., Schmitz, R. W., Krainitzki, H. & Paabo, S. (1999). DNA sequence of the mitochondrial hypervariable region II from the Neandertal type specimen. *Proceedings of the National Academy of Sciences, USA*, **96**, 5581–5585.

Krings, M., Stone, A., Scmitz, R. W., Krainitzki, H., Stoneking, M. & Paabo, S. (1997). Neandertal DNA sequences and the origin of modern humans. *Cell*, **90**, 19–30.

Kuhn, S. L. (1995). *Mousterian Lithic Technology. An Ecological Perspective*. Princeton, Princeton University Press.

Kuhn, S. L. & Stiner, M. C. (1998). The earliest Aurignacian of Riparo Mochi (Liguria, Italy). *Current Anthropology*, **39**, S175–S189.

Kuhn, S. L., Stiner, M. C., Reese, D. S. & Gulec, E. (2001). Ornaments of the earliest Upper Palaeolithic: new insights from the Levant. *Proceedings of the National Academy of Sciences, USA*, **98**, 7641–7646.

Kuhn, T. S. (1970). *The Structure of Scientific Revolutions*. Chicago, Chicago University Press.

Kukla, G. (1997). Was the Holocene climate uniquely benign or is the Eemian catching cold from the reinforcement syndrome virus? *NATO ASI Series*, **I 49**, 699–709.

Kukla, G. J. (2000). The last interglacial. *Science*, **287**, 987–988.

Kukla, G. J., Bender, M. L., de Beaulieu, J-L. *et al.* (2002). Last interglacial climates. *Quaternary Research*, **58**, 2–13.

Kukla, G., McManus, J. F., Rousseau, D-D. & Chuine, I. (1997). How long and how stable was the last interglacial? *Quaternary Science Reviews*, **16**, 605–612.

Lack, D. (1971). *Ecological Isolation in Birds*. Oxford, Blackwell.

Lahr, M. M. (1994). The Multiregional model of modern human origins: a reassessment of its morphological basis. *Journal of Human Evolution*, **26**, 23–56.

Lahr, M. & Foley, R. (1994). Multiple dispersals and modern human origins. *Evolutionary Anthropology*, **3**, 48–60.

Lahr, M. M. & Foley, R. (1998). Towards a theory of modern human origins: geography, demography, and diversity in recent human evolution. *Yearbook of Physical Anthropology*, **41**, 137–176.

Lahr, M. M. & Wright, R. V. S. (1996). The question of robusticity and the relationship between cranial size and shape in *Homo sapiens*. *Journal of Human Evolution*, **31**, 157–191.

Lai, C. S. L., Fisher, S. E., Hurst, J. A., Vargha-Khadem, F. & Monaco, A. P. (2001). A forkhead-domain gene is mutated in a severe speech and language disorder. *Nature*, **413**, 519–523.

Lalueza Fox, C. & Pérez-Pérez, A. (1993). The diet of the Neanderthal Child Gibraltar 2 (Devil's Tower) through the study of the vestibular striation pattern. *Journal of Human Evolution*, **24**, 29–41.

Lalueza, C., Pérez-Pérez, A. & Turbón, D. (1996). Dietary inferences through buccal microwear analysis of Middle and Upper Pleistocene human fossils. *American Journal of Physical Anthropology*, **100**, 367–387.

Lambeck, K. & Chappell, J. (2001). Sea level change through the last glacial cycle. *Science*, **292**, 679–686.

Lambeck, K., Esat, T. M. & Potter, E-K. (2002a). Links between climate and sea levels for the past three million years. *Nature*, **419**, 199–206.

Lambeck, K., Yokoyama, Y. & Purcell, T. (2002b). Into and out of the Last Glacial Maximum: sea-level change during Oxygen Isotope Stages 3 and 2. *Quaternary Science Reviews*, **21**, 343–360.

Laville, H., Rigaud, J-P. & Sachett, J. (1980). *Rock Shelters of the Périgord: Geological stratigraphy and archaeological succession.* New York, Academic Press.

Lebel, S., Trinkaus, E., Faure, M. *et al.* (2001). Comparative morphology and paleobiology of Middle Pleistocene human remains from the Bau de l'Aubesier, Vaucluse, France. *Proceedings of the National Academy of Sciences, USA,* **98**, 11097–11102.

Lee, R. B. (1979). *The !Kung San. Men, Women, and Work in a Foraging Society.* Cambridge: Cambridge University Press.

Lehman, C. L. & Tilman, D. (1997). Competition in spatial habitats. In *Spatial Ecology. The Role of Space in Population Dynamics and Interspecific Interactions,* ed. D. Tilman & P. Karieva, pp. 185–203. Princeton, Princeton University Press.

Levin, S. A. (1992). The problem of pattern and scale in ecology: the Robert MacArthur Award Lecture. *Ecology,* **73**, 1943–1967.

Levine, M. A. (1999). Botai and the origins of horse domestication. *Journal of Anthropological Archaeology,* **18**, 29–78.

Levins, R. (1968). *Evolution in Changing Environments.* Princeton, Princeton University Press.

Lev-Yadun, S. (2000). The cradle of agriculture. *Science,* **288**, 1602–1603.

Lewis, M. A. (1997). Variability, patchiness, and jump dispersal in the spread of an invading population. In *Spatial Ecology. The Role of Space in Population Dynamics and Interspecific Interactions,* ed. D. Tilman & P. Karieva, pp. 46–74. Princeton, Princeton University Press.

Lezine, A-M., Turon, J-L. & Buchet, G. (1995). Pollen analyses off Senegal: evolution of the coastal palaeoenvironment during the last deglaciation. *Journal of Quaternary Science,* **10**, 95–105.

Lieberman, D. E. (1998). Neandertal and early Modern Human mobility patterns: comparing archaeological and anatomical evidence. In *Neandertals and Modern Humans in Western Asia,* ed. T. Akazawa, K. Aoki & O. Bar-Yosef, pp. 263–276. New York, Plenum Press.

Lieberman, D. E., McBratney, B. M. & Krovitz, G. (2002). The evolution and development of cranial form in *Homo sapiens. Proceedings of the National Academy of Sciences, USA,* **99**, 1134–1139.

Lieberman, D. E. & Shea, J. J. (1994). Behavioral differences between archaic and modern humans in the Levantine Mousterian. *American Anthropologist,* **96**, 300–332.

Lieberman, P. (1989). The origins of some aspects of human language and cognition. In *The Human Revolution. Behavioural and Biological Perspectives in the Origins of Modern Humans,* ed. C. Stringer & P. Mellars, pp. 391–414. Edinburgh, Edinburgh University Press.

Lindly, J. M. & Clark, G. A. (1990). Symbolism and modern human origins. *Current Anthropology,* **31**, 233–261.

Lindstedt, S. L. & Boyce, M. S. (1985). Seasonality, fasting endurance, and body size in mammals. *American Naturalist,* **125**, 873–878.

Lister, A. M. (1996). Evolution and taxonomy of Eurasian mammoths. In *The Proboscidea. Evolution and Palaeoecology of Elephants and Their Relatives,* ed. J. Shoshani & P. Tassy, pp. 203–213. Oxford, Oxford University Press.

Lynch-Stieglitz, J., Curry, W. B. & Slowey, N. (1999). Weaker Gulf Stream in the Florida Straits during the Last Glacial Maximum. *Nature*, **402**, 644–648.

Maarleveld, G. C. (1976). Periglacial phenomena and the mean annual temperature during the last glacial time in the Netherlands. *Biuletyn Peryglacjalny*, **26**, 57–78.

MacArthur, R. H. (1984). *Geographical Ecology. Patterns in the Distribution of Species.* Princeton, Princeton University Press.

MacArthur, R. H. & Wilson, E. O. (1967). *The Theory of Island Biogeography.* Princeton, Princeton University Press.

Macaulay, V., Richards, M., Hickey, E. *et al.* (1999). The emerging tree of west Eurasian mtDNAs: a synthesis of control-region sequences and RFLPs. *American Journal of Human Genetics*, **64**, 232–249.

MacFadden, B. J. (1992). *Fossil Horses. Systematics, Paleobiology, and Evolution of the Family Equidae.* Cambridge, Cambridge University Press.

Madsen, M., Lipo, C. & Cannon, M. (1999). Fitness and reproductive trade-offs in uncertain environments: explaining the evolution of cultural elaboration. *Journal of Anthropological Archaeology*, **18**, 251–281.

Malaspina, P., Cruciani, F., Ciminelli, B. M. *et al.* (1998). Network analyses of Y-chromosomal types in Europe, northern Africa, and western Asia reveal specific patterns of geographic distribution. *American Journal of Human Genetics*, **63**, 847–860.

Manzi, G., Mallegni, F. & Ascenzi, A. (2001). A cranium for the earliest Europeans: Phylogenetic position of the hominid from Ceprano, Italy. *Proceedings of the National Academy of Sciences, USA*, **98**, 10011–10016.

Mareau, C. W. (1998). A critique of the evidence for scavenging by Neanderthals and early modern humans: new data from Kobeh Cave (Zagros Mountains, Iran) and Die Kelders Cave I Layer 10 (South Africa). *Journal of Human Evolution*, **35**, 111–136.

Mareau, C. W. & Kim, S. Y. (1998). Mousterian large-mammal remains from Kobeh Cave: behavioural implications for Neanderthals and early modern humans. *Current Anthropology*, **39**, S79–S113.

Markova, A. K., Simakova, A. N., Puzachenko, A. Y. & Kitaev, L. M. (2002). Environments of the Russian Plain during the Middle Valdai Briansk Interstade (33,000–24,000 yr BP) indicated by fossil mammals and plants. *Quaternary Research*, **57**, 391–400.

Marks, A. E. (1992). Upper Pleistocene archaeology and the origins of modern man: a view from the Levant and adjacent areas. In *The Evolution and Dispersal of Modern Humans in Asia*, ed. T. Akazawa, K. Aoki & T. Kimura, pp. 229–251. Tokyo, Hokusen-sha.

Marks, A. E. (2000). Upper Paleolithic occupation of Portugal: residents vs. visitors. In *Paleolítico da Península Ibérica*, Actas do 3. o Congresso de Arqueologia Peninsular, **2**, 341–350.

Maroto, J., Soler, N. & Fullola, J. M. (1996). Cultural change between Middle and Upper Palaeolithic in Catalonia. In *The Last Neanderthals, The First Anatomically Modern Humans*, ed. E. Carbonell & M. Vaquero, pp. 219–250. Tarragona, Universitat Rovira i Virgili.

Marth, G., Schuler, G., Yeh, R. *et al.* (2003). Sequence variations in the public genome data reflect a bottlenecked population history. *Proceedings of the National Academy of Sciences, USA*, **100**, 376–381.

Martin, P. S. & Klein, R. G. (Eds) (1984). *Quaternary Extinctions. A Prehistoric Revolution.* Tucson, University of Arizona Press.

Martínez Navarro, B., Turq, A., Agustí Ballester, J. & Oms, O. (1997). Fuente Nueva-3 (Orce, Granada, Spain) and the first human occupation of Europe. *Journal of Human Evolution*, **33**, 611–620.

May, R. (1973). *Stability and Complexity in Model Ecosystems.* Princeton, Princeton University Press.

Maynard Smith, J. (1966). Sympatric speciation. *American Naturalist*, **100**, 637–650.

Mayr, E. (1942). *Systematics and the Origin of Species.* New York, Dover.

Mayr, E. (1950). Taxonomic categories of fossil hominids. *Cold Spring Harbor Symposium on Quantitative Biology* **15**, 109–118.

Mayr, E. (1963). *Animal Species and Evolution.* Cambridge, MA, Harvard University Press.

McBrearty, S. & Brooks, A. S. (2000). The revolution that wasn't: a new interpretation of the origin of modern human behaviour. *Journal of Human Evolution*, **39**, 453–563.

McGlone, M. (1996). When history matters: scale, time, climate and tree diversity. *Global Ecology and Biogeography Letters*, **5**, 309–314.

McHenry, H. M. (1994). Behavioral ecological implications of early hominid body size. *Journal of Human Evolution*, **27**, 77–87.

McHenry, H. M. & Berger, L. R. (1998). Body proportions in *Australopithecus afarensis* and *A. africanus* and the origin of the genus *Homo*. *Journal of Human Evolution*, **35**, 1–22.

McNab, B. K. (1971). On the ecological significance of Bergmann's Rule. *Ecology*, **52**, 845–854.

Mellars, P. (1996). *The Neanderthal Legacy. An Archaeological Perspective from Western Europe.* Princeton, Princeton University Press.

Mellars, P. (1999). The Neanderthal problem continued. *Current. Anthropology*, **40**, 341–364.

Miller, K. G., Fairbanks, R. G. & Mountain, G. S. (1987). Tertiary oxygen isotope synthesis, sea level history, and continental margin erosion. *Paleoceanography*, **2**, 1–19.

Milton, K. (1981). Distribution patterns of tropical plant foods as an evolutionary stimulus to primate mental development. *American Anthropologist*, **83**, 534–548.

Mithen, S. (1996). *The Prehistory of the Mind. A Search for the Origins of Art, Religion and Science.* London, Thames & Hudson.

Mithen, S. & Reed, M. (2002). Stepping out: a computer simulation of hominid dispersal from Africa. *Journal of Human Evolution*, **43**, 433–462.

Miyamoto, M. M., Kraus, F. & Ryder, O. A. (1990). Phylogeny and evolution of antlered deer determined from mitochondrial DNA sequences. *Proceedings of the National Academy of Sciences, USA*, **87**, 6127–6131.

Montet-White, A. (1994). Alternative interpretations of the late Upper Palaeolithic in Central Europe. *Annual Review of Anthropology*, **23**, 483–508.

Montuire, S. & Marcolini, F. (2002). Palaeoenvironmental significance of the mammalian faunas of Italy since the Pliocene. *Journal of Quaternary Science*, **17**, 87–96.

Mussi, M. (1999). The Neanderthals in Italy: a tale of many caves. In *The Middle Palaeolithic Occupation of Europe*, ed. W. Roebroeks & C. Gamble, pp. 49–80. Leiden, University of Leiden Press.

Mussi, M. & Roebroeks, W. (1996). The big mosaic. *Current Anthropology*, **37**, 697–699.

Narcisi, B. (2001). Palaeoenvironmental and palaeoclimatic implications of the Late-Quaternary sediment record of Vico volcanic lake (central Italy). *Journal of Quaternary Science*, **16**, 245–255.

Nicholas, G. P. (1998). Wetlands and hunter–gatherers: a global perspective. *Current Anthropology*, **39**, 720–731.

Nikaido, M., Rooney, A. P. & Okada, N. (1999). Phylogenetic relationships among cetartiodactyls based on insertions of short and long interspersed elements: hippopotamuses are the closest extant relatives of whales. *Proceedings of the National Academy of Sciences, USA*, **96**, 10261–10266.

Nimmergut, A. P., Allen, J. R. M., Jones, V. J., Huntley, B. & Battarbee, R. W. (1999). Submillennial environmental fluctuations during marine Oxygen Isotope Stage 2: a comparative analysis of diatom and pollen evidence from Lago Grande di Monticchio, South Italy. *Journal of Quaternary Science*, **14**, 111–123.

Nordborg, M. (1998). On the probability of Neanderthal ancestry. *American Journal of Human Genetics*, **63**, 1237–1240.

O'Brien, E. M. & Peters, C. R. (1999). Landforms, climate, ecogeographic mosaics, and the potential for hominid diversity in Pliocene Africa. In *African Biogeography, Climate Change, and Human Evolution*, ed. T. G. Bromage & F. Schrenk, pp. 115–137. Oxford, Oxford University Press.

O'Connell, J. F. & Hawkes, K. (1988). Hadza hunting, butchering, and bone transport and their archaeological implications. *Journal of Anthropological Research*, **44**, 113–161.

Ohnuma, K., Aoki, K. & Akazawa, T. (1997). Transmission of tool-making through verbal and non-verbal communication: preliminary experiments in Levallois flake production. *Anthropological Science*, **105**, 159–168.

Oliver, B. M. & Juan-Cabanilles, J. (2000). Epipaleolíticos y Neolíticos en la Península Ibérica del VII al V milenio a. de c. Grupos, territorios y procesos culturales. In *El Paisaje en el Neolítico Mediterraneo*, pp. 12–25. Valencia, Universidad de Valencia.

Oms, O., Parés, J. M., Martínez-Navarro, B. *et al.* (2000). Early human occupation of Western Europe: paleomagnetic dates for two paleolithic sites in Spain. *Proceedings of the National Academy of Sciences, USA*, **97**, 10666–10670.

Oppo, D. W., McManus, J. F. & Cullen, J. L. (1998). Abrupt climate events 500,000 to 340,000 years ago: evidence from subpolar North Atlantic sediments. *Science*, **279**, 1335–1338.

O'Regan, H. J., Turner, A. & Wilkinson, D. M. (2002). European Quaternary refugia: a factor in large carnivore extinction? *Journal of Quaternary Science*, **17**, 789–795.

Otte, M. (1994). Origine de l'homme moderne: approche comportmentale. *C. R. Academie Sciences Paris*, **318**, 267–273.

Ovchinnikov, I. V., Gotherstrom, A., Romanova, G. P., Kharitonov, V. M., Liden, K. & Goodwin, W. (2000). Molecular analysis of Neanderthal DNA from the northern Caucasus. *Nature*, **404**, 490–493.

Owen-Smith, R. N. (1988). *Megaherbivores. The Influence of very Large Body Size on Ecology*. Cambridge, Cambridge University Press.

Parés, J. M. & Pérez-González, A. (1995). Paleomagnetic age for hominid fossils at Atapuerca Archaeological site, Spain. *Science*, **269**, 830–832.

Parés, J. M. & Pérez-González, A. (1999). Magnetochronology and stratigraphy at Gran Dolina section, Atapuerca (Burgos, Spain). *Journal of Human Evolution*, **37**, 325–342.

Park, R. W. (1998). On the Dorset/Thule analogy for the Middle/Upper Paleolithic transition. *Current Anthropology*, **39**, 355–356.

Parsons, P. A. (1993). Evolutionary adaptation and stress: energy budgets and habitats preferred. *Behavior Genetics*, **23**, 231–238.

Pavlov, P., Svendsen, J. I. & Indrelid, S. (2001). Human presence in the European Arctic nearly 40,000 years ago. *Nature*, **413**, 64–67.

Pearson, O. (2000). Activity, climate, and postcranial robusticity. *Current Anthropology*, **41**, 569–607.

Pease, C. M., Lande, R. & Bull, J. J. (1989). A model of population growth, dispersal and evolution in a changing environment. *Ecology*, **70**, 1657–1664.

Peteet, D. (2000). Sensitivity and rapidity of vegetational response to abrupt climate change. *Proceedings of the National Academy of Sciences, USA*, **97**, 1359–1361.

Peters, R. H. (1983). *The Ecological Implications of Body Size*. Cambridge, Cambridge University Press.

Pickford, M. & Morales, J. (1994). Biostratigraphy and palaeobiogeography of East Africa and the Iberian Peninsula. *Palaeogeography, Palaeoclimatology, Palaeoecology*, **112**, 297–322.

Ponce de León, M. S. & Zollikofer, C. P. E. (2001). Neanderthal cranial ontogeny and its implications for late hominid diversity. *Nature*, **412**, 534–538.

Pond, C. M. (1978). Morphological aspects and the ecological and mechanical consequences of fat deposition in wild vertebrates. *Annual Review of Ecology and Systematics*, **9**, 519–570.

Pond, C. (1999). *The Fats of Life*. Cambridge, Cambridge University Press.

Pons, A. (1984). Les changements de la végetation de la region méditerrané Durant le Pliocene et le Quaternaire en relation avec l'histoire du climat et de l'action de l'homme. *Webbia*, **38**, 427–439.

Pope, G. G. (1983). Evidence for the age of the Asian Hominidae. *Proceedings of the National Academy of Sciences, USA*, **80**, 4988–4992.

Porat, N., Chazan, M., Schwarcz, H. & Kolska Horwitz, L. (2002). Timing of the Lower to Middle Paleolithic boundary: new dates from the Levant. *Journal of Human Evolution*, **43**, 107–122.

Porcasi, J. F., Jones, T. L. & Raab, L. M. (2000). Trans-Holocene marine mammal exploitation on San Clemente Island, California: a tragedy of the commons revisited. *Journal of Anthropological Archaeology*, **19**, 200–220.

Porter, A. H. & Johnson, N. A. (2002). Speciation despite gene flow when developmental pathways evolve. *Evolution*, **56**, 2103–2111.

Potts, R. (1996a). Evolution and climate variability. *Science*, **273**, 922–923.

Potts, R. (1996b). *Humanity's Descent. The Consequences of Ecological Instability.* New York, William Morrow.

Potts, R. (1998). Variability selection in hominid evolution. *Evolutionary Anthropology,* **7**, 81–96.

Prins, H. H. T. & Olff, H. (1998). Species-richness of African grazer assemblages: towards a functional explanation. In *Dynamics of Tropical Communities,* ed. D. M. Newbery, H. H. T. Prins & N. Brown, pp. 449–490. Oxford, Blackwell Science.

Prins, H. H. T. & Reitsma, J. M. (1989). Mammalian biomass in an African equatorial rain forest. *Journal of Animal Ecology,* **58**, 851–861.

Prokopenko, A. A., Karabanov, E. B., Williams, D. F. & Khursevich, G. K. (2002). The stability and the abrupt ending of the last interglaciation in southeastern Siberia. *Quaternary Research,* **58**, 56–59.

Pulliam, H. R., Dunning, J. B. & Liu, J. (1992). Population dynamics in complex landscapes: a case study. *Ecological Applications,* **2**, 165–177.

Queiroz do Amaral, L. (1996). Loss of body hair, bipedality and thermoregulation. Comments on recent papers in the *Journal of Human Evolution. Journal of Human Evolution,* **30**, 357–366.

Quintana-Murci, L., Semino, O., Bandelt, H-J., Passarino, G., McElreavey, K. & Santachiara-Benerecetti, A. S. (1999). Genetic evidence of an early exit of *Homo sapiens sapiens* through eastern Africa. *Nature Genetics,* **23**, 437–441.

Rak, Y. (1990). On the differences between two pelvises of Mousterian context from the Qafzeh and Kebara caves, Israel. *American Journal of Physical Anthropology,* **81**, 323–332.

Raposo, L. (2000). The Middle–Upper Palaeolithic Transition in Portugal. In *Neanderthals on the Edge,* ed. C. B. Stringer, R. N. E. Barton, & J. C. Finlayson, pp. 95–105. Oxford, Oxbow Books.

Raposo, L. & Cardoso, J. L. (1998). Las industrias liticas de la gruta de Columbeira (Bombarral, Portugal) en el contexto del musteriense final de la Peninsula Ibérica. *Trabajos de Prehistoria,* **55**, 39–62.

Raynal, J. P., Sbihi Alaoui, F. Z., Geraads, D., Magoga, L. & Mohi, A. (2001). The earliest occupation of North-Africa: the Moroccan perspective. *Quaternary International,* **75**, 65–75.

Reed, K. E. (1997). Early hominid evolution and ecological change through the African Plio-Pleistocene. *Journal of Human Evolution,* **32**, 289–322.

Reich, D. E., Cargill, M., Bolk, S. *et al.* (2001). Linkage disequilibrium in the human genome. *Nature,* **411**, 199–204.

Reille, M. (1984). Origine de la vegetation actuelle de la Corse sudorientale; analyse pollinique de cinq marias cotiers. *Pollen et Spores,* **26**, 43–60.

Reille, M., Andrieu, V., de Beaulieu, J-L., Guenet, P. & Goeury, C. (1998). A long pollen record from Lac du Bouchet, Massif Central, France: for the period ca. 325 to 100 ka BP (OIS 9c to OIS 5e). *Quaternary Science Reviews,* **17**, 1107–1123.

Reille, M. & de Beaulieu, J-L. (1995). Long Pleistocene pollen records from the Praclaux Crater, south-central France. *Quaternary Research,* **44**, 205–215.

Reille, M., de Beaulieu, J-L., Svoboda, H., Andrieu-Ponel, V. & Goeury, C. (2000). Pollen analytical biostratigraphy of the last five climatic cycles from a long continental sequence from the Velay region (Massif Central, France). *Journal of Quaternary Science,* **15**, 665–685.

Reletheford, J. H. (1999). Models, predictions, and the fossil record of modern human origins. *Evolutionary Anthropology*, **8**, 7–10.

Reletheford, J. H. & Jorde, L. B. (1999). Genetic evidence for larger African population size during recent human evolution. *American Journal of Physical Anthropology*, **108**, 251–260.

Rice, W. R. & Hostert, E. E. (1993). Laboratory experiments on speciation: what have we learned in 40 years? *Evolution*, **47**, 1637–1653.

Richards, M., Macaulay, V., Hickey, E. *et al.* (2000). Tracing European founder lineages in the Near Eastern mtDNA Pool. *American Journal of Human Genetics*, **67**, 1251–1276.

Richards, M. B. & Macaulay, V. A. (2001). The mitochondrial gene tree comes of age. *American Journal of Human Genetics*, **68**, 1315–1320.

Richards, M. B., Macaulay, V. A., Bandelt, H.-J. & Sykes, B. C. (1998). Phylogeography of mitochondrial DNA in western Europe. *Annals of Human Genetics*, **62**, 241–260.

Richards, M. P., Pettitt, P. B., Stiner, M. C. & Trinkaus, E. (2001). Stable isotope evidence for increasing dietary breadth in the European mid-Upper Paleolithic. *Proceedings of the National Academy of Sciences, USA*, **98**, 6528–6532.

Richards, M. P., Pettitt, P. B., Trinkaus, E., Smith, F. H., Paunovic, M. & Karavanic, I. (2000). Neanderthal diet at Vindija and Neanderthal predation: the evidence from stable isotopes. *Proceedings of the National Academy of Sciences, USA*, **97**, 7663–7666.

Richmond, B. G., Begun, D. R. & Strait, D. S. (2001). Origin of human bipedalism: the knuckle-walking hypothesis revisited. *Yearbook of Physical Anthropology*, **44**, 70–105.

Ricklefs, R. E. & Miles, D. B. (1994). Ecological and evolutionary inferences from morphology: an ecological perspective. In *Ecological Morphology. Integrative Organismal Biology*, ed. P. C. Wainwright & S. M. Reilly, pp. 13–41. Chicago, University of Chicago Press.

Riel-Salvatore, J. & Clark, G. A. (2001). Grave markers. Middle and Upper Paleolithic burials and the use of chronotypology in contemporary Paleolithic research. *Current Anthropology*, **42**, 449–479.

Riolo, R. L., Cohen, M. D. & Axelrod, R. (2001). Evolution of cooperation without reciprocity. *Nature*, **414**, 441–443.

Ripoll López, S. & Cacho Quesada, C. (1990). Le Solutréen dans le sud de la Peninsule Iberique. Feuilles de pierre. Les industries à pointes foliacées du Paléolithique supérieur européen, Krakòw 1989. *Eraul*, **42**, 449–465.

Rivas-Martínez, S. (1981). Les étages bioclimatiques de la vegetation de la Péninsule Ibérique. *Actas III Congreso Optima. Anales Jardin Botanico Madrid*, **37**, 251–268.

Rivas-Martínez, S. (1987). Memoria del mapa de series de vegetación de España. Madrid, ICONA.

Rivas-Martínez, S. (1996). *Bioclimatic Map of Europe*. León, Universidad de León.

Roberts, G. & Sherratt, T. N. (2002). Does similarity breed cooperation? *Nature*, **418**, 499–500.

Roebroeks, W., Conrad, N. J. & van Kolfschoten, T. (1992). Dense forests, cold steppes, and the Palaeolithic settlement of northern Europe. *Current Anthropology*, **33**, 551–586.

Rogers, A. R. & Jorde, L. (1995). Genetic evidence on modern human origins. *Human Biology*, **67**, 1–36.

Rohling, E. J., Fenton, M., Jorissen, F. J., Bertrand, P., Ganssen, G. & Caulet, J. P. (1998). Magnitudes of sea-level lowstands of the past 500,000 years. *Nature*, **394**, 162–165.

Rolland, N. (1998). Comment to Van Peer: the nile corridor and the Out-of-Africa Model. *Current Anthropology*, **39** (Suppl.), S133–S134.

Rolland, N. & Dibble, H. L. (1990). A new synthesis of Middle Paleolithic variability. *American Antiquity*, **55**, 480–499.

Ronen, A. (1992). The emergence of blade technology: cultural affinities. In *The Evolution and Dispersal of Modern Humans in Asia*, ed. T. Akazawa, K. Aoki & T. Kimura, pp. 217–228. Tokyo, Hokusen-sha.

Rose, J., Meng, X. & Watson, C. (1999). Palaeoclimate and palaeoenvironmental responses in the western Mediterranean over the last 140 ka: evidence from Mallorca, Spain. *Journal of the Geological Society, London*, **156**, 435–448.

Rosenberg, K. R. (1988). The functional significance of Neandertal pubic length. *Current Anthropology*, **29**, 595–617.

Rosenberg, M. (1998). Cheating at musical chairs: territoriality and sedentism in an evolutionary context. *Current Anthropology*, **39**, 653–681.

Rosenzweig, M. L. (1968). The strategy of body size in mammalian carnivores. *American Midland Naturalist*, **80**, 299–315.

Roucoux, K. H., Shackleton, N. J., de Abreu, L., Schonfeld, J. & Tzedakis, P. C. (2001). Combined marine proxy and pollen analyses reveal rapid Iberian vegetation response to North Atlantic millennial-scale climate oscillations. *Quaternary Research*, **56**, 128–132.

Rousseau, D-D., Gerasimenko, N., Matviischina, Z. & Kukla, G. (2001). Late Pleistocene environments of the Central Ukraine. *Quaternary Research*, **56**, 349–356.

Ruddiman, W. F. & McIntyre, A. (1977). Late quaternary surface ocean kinematics and climate change in the high-latitude North Atlantic. *Journal of Geophysical Research*, **82**, 3877–3887.

Ruddiman, W. F., McIntyre, A. & Raymo, M. (1986). Palaeoenvironmental results from North Atlantic Sites 607 and 609. *Initial Reports of the Deep-Sea Drilling Project*, **94**, 855–878. Washington DC, US Govt. Printing Office.

Ruff, C. B. (1994). Morphological adaptation to climate in modern and fossil hominids. *Yearbook of Physical Anthropology*, **37**, 65–107.

Russell, M. D. (1985). The supraorbital torus: 'A most remarkable peculiarity'. *Current Anthropology*, **26**, 337–360.

Salzmann, U., Hoelzmann, P. & Morczinek, I. (2002). Late Quaternary climate and vegetation of the Sudanian zone of northeast Nigeria. *Quaternary Research*, **58**, 73–83.

Sánchez Goñi, M. F., Eynaud, F., Turon, J-L. & Shackleton, N. J. (1999). High resolution palynological record off the Iberian margin: direct land-sea correlation for the last interglacial complex. *Earth and Planetary Science Letters*, **171**, 123–137.

Sánchez Goñi, M. F., Turon, J-L., Eynaud, F. & Gendreau, S. (2000a). European climatic response to millennial-scale changes in the atmosphere-ocean system during the last glacial period. *Quaternary Research*, **54**, 394–403.

Sánchez Goñi, M. F., Turon, J-L., Eynaud, F., Shackleton, N. J. & Cayre, O. (2000b). Direct land/sea correlation of the Eemian, and its comparison with the Holocene: a high resolution palynological record off the Iberian margin. *Netherlands Journal of Geosciences*, **79**, 345–354.

Saunders, J. J. (1996). North American Mammutidae. In *The Proboscidea. Evolution and Palaeoecology of Elephants and Their Relatives*, ed. J. Shoshani & P. Tassy, pp. 271–279. Oxford, Oxford University Press.

Savage-Rumbaugh, S. & Lewin, R. (1994). *Kanzi. The Ape at the Brink of the Human Mind*. New York, John Wiley & Sons.

Schluter, D. & Nagel, L. M. (1995). Parallel Speciation by Natural Selection. *American Naturalist*, **146**, 292–301.

Schluter, D. & Price, T. (1993). Honesty, perception and population divergence in sexually selected traits. *Proceedings of the Royal Society London, Series B*, **253**, 117–122.

Scholz, M., Bachmann, L., Nicholson, G. J. *et al.* (2000). Genomic differentiation of Neanderthals and anatomically modern man allows a fossil-DNA-based classification of morphologically indistinguishable hominid bones. *American Journal of Human Genetics*, **66**, 1927–1932.

Schrag, D. P., Hampt, G. & Murray, D. W. (1996). Pore fluid constraints on the temperature and oxygen isotopic composition of the glacial ocean. *Science*, **272**, 1930–1932.

Semino, O., Passarino, G., Oefner, P. J. *et al.* (2000). The Genetic Legacy of Paleolithic *Homo sapiens sapiens* in Extant Europeans: A Y Chromosome Perspective. *Science*, **290**, 1155–1159.

Shackleton, N. J., Backman, J., Zimmerman, H. *et al.* (1984). Oxygen isotope calibration of the onset of ice-rafting and history of glaciation in the North Atlantic region. *Nature*, **307**, 620–623.

Shackleton, N. J., Chapman, M., Sánchez-Goñi, M. F., Pailler, D. & Lancelot, Y. (2002). The classic marine isotope substage 5e. *Quaternary Research*, **58**, 14–16.

Shackleton, N. J. & Opdyke, N. D. (1973). Oxygen isotope and palaeomagnetic stratigraphy of equatorial Pacific core V28–238: oxygen isotope temperatures and ice volumes on a 10^5 and 10^6 year scale. *Quaternary Research*, **3**, 39–55.

Shackleton, N. J. & Opdyke, N. D. (1976). Oxygen isotope and paleomagnetic stratigraphy of Pacific core V28–239, late Pliocene to latest Pleistocene. *Geological Society of America Memoir*, **145**, 449–464.

Shea, J. J. (1998). Neandertal and early modern human behavioral variability. *Current Anthropology*, **39**, S45-S78.

Shen, P., Wang, F., Underhill, P. A. *et al.* (2000). Population genetic implications from sequence variation in four chromosome genes. *Proceedings of the National Academy of Sciences, USA*, **97**, 7354–7359.

Sherry, S. T., Rogers, A. R., Harpending, H., Soodyall, H., Jenkins, T. & Stoneking, M. (1994). Mismatch distributions of mtDNA reveal recent human population expansions. *Human Biology*, **66**, 761–775.

Shoshani, J., West, R. M., Court, N., Savage, R. J. G. & Harris, J. M. (1996). The earliest proboscideans: general plan, taxonomy, and palaeoecology. In *The Proboscidea.*

Evolution and Palaeoecology of Elephants and Their Relatives, ed. J. Shoshani & P. Tassy, pp. 57–75. Oxford, Oxford University Press.

Siguroardottir, S., Helgason, A., Gulcher, J. R., Stefanson, K. & Donnelly, P. (2000). The mutation rate in the human mtDNA control region. *American Journal of Human Genetics*, **66**, 1599–1609.

Simek, J. & Snyder, L. (1988). Changing assemblage diversity in Périgord archaeofaunas. In *The Upper Pleistocene prehistory of western Eurasia*, ed. H. Dibble & A. Montet-White, pp. 321–332. Philadelphia, University Museum Press.

Simmons, T. & Smith, F. H. (1991). Human population relationships in the late Pleistocene. *Current Anthropology*, **32**, 623–627.

Simpson, G. G. (1953). The Baldwin effect. *Evolution*, **7**, 110–117.

Simpson, G. G. (1951). The species concept. *Evolution*, **5**, 285–298.

Singleton, I. & van Schaik, C. P. (2001). Orangutan home range size and its determinants in a Sumatran swamp forest. *International Journal of Primatology*, **22**, 877–911.

Smith, F. H. (1983). A behavioural interpretation of changes in craniofacial morphology across the archaic/modern *Homo sapiens* transition. In *The Mousterian Legacy*, ed. E. Trinkaus, pp. 141–163. British Archaeological Reports, S164.

Smith, F. H. (1992). The role of continuity in modern human origins. In *Continuity or Replacement. Controversies in* Homo sapiens *Evolution*, ed. G. Brauer & F. H. Smith, pp. 145–156. Rotterdam, A. A. Balkema.

Smith, F. H., Falsetti, A. B. & Simmons, T. (1995). Circum-Mediterranean biological connections and the pattern of late Pleistocene human evolution. In *Man and environment in the Palaeolithic*, ed. H. Ullrich, pp. 197–208. Liege, Etudes et Recherches Archaeologiques de l'Université de Liege **62**.

Smith, F. H., Trinkaus, E., Pettitt, P. B., Karavanic, I. & Paunovic, M. (1999). Direct radiocarbon dates for Vindija G_1 and Velika Pecina Late Pleistocene hominid remains. *Proceedings of the National Academy of Sciences, USA*, **96**, 12281–12286.

Soffer, O. (1985). *The Upper Paleolithic of the Central Russian Plain*. San Diego, Academic Press.

Soffer, O. (1989). The Middle to Upper Palaeolithic transition on the Russian Plain. In *The Human Revolution. Behavioural and biological perspectives on the origins of modern humans*, ed., C. Stringer & P. Mellars, pp. 714–742. Edinburgh, Edinburgh University Press.

Soffer, O. (1994). Ancestral lifeways in Eurasia – the Middle and Upper Palaeolithic records. In *Origins of Anatomically Modern Humans*, ed. M. H. Nitecki, & D. V. Nitecki, pp. 101–119. New York, Plenum Press.

Sokal, R. R., Oden, N. L., Walker, J. & Waddle, D. M. (1997). Using distance matrices to choose between competing theories and an application to the origin of modern humans. *Journal of Human Evolution*, **32**, 501–522.

Sokal, R. R. & Rohlf, F. J. (1981). *Biometry*. 2nd edn. New York, Freeman.

Speth, J. D. (1989). Early hominid hunting and scavenging: the role of meat as an energy source. *Journal of Human Evolution*, **18**, 329–343.

Speth, J. D. (1991). Protein selection and avoidance strategies of contemporary and ancestral foragers: unresolved issues. *Philosophical Transactions of the Royal Society, London, Series B*, **334**, 265–270.

Srikosamatara, S. (1993). Density and biomass of large herbivores and other mammals in a dry tropical forest, western Thailand. *Journal of Tropical Ecology*, **9**, 33–43.

Stanford, C. B. (1999). *The Hunting Apes. Meat Eating and the Origin of Human Behavior*. Princeton, Princeton University Press.

Stanford, C. B. & Bunn, H. T., ed. (2001). *Meat-eating and Human Evolution*. Oxford, Oxford University Press.

Stanley, H. F., Kadwell, M. & Wheeler, J. C. (1994). Molecular evolution of the family Camelidae: a mitochondrial DNA study. *Proceedings of the Royal Society London, Series B*, **256**, 1–6.

Sterling, S. (1999). Mortality Profiles as Indicators of Slowed Reproductive Rates: Evidence from Ancient Egypt. *Journal of Anthropological Archaeology*, **18**, 319–343.

Stewart, J. R. & Lister, A. M. (2001). Cryptic northern refugia and the origins of the modern biota. *Trends in Ecology and Evolution*, **16**, 608–613.

Stiner, M. (1994). *Honor among Thieves. A Zooarchaeological Study of Neandertal Ecology*. Princeton, Princeton University Press.

Stiner, M. C. (2001). Thirty years on the 'Broad Spectrum Revolution' and paleolithic demography. *Proceedings of the National Academy of Sciences, USA*, **98**, 6993–6996.

Stiner, M. C. & Kuhn, S. L. (1992). Subsistence, technology, and adaptive variation in middle Paleolithic Italy. *American Anthropologist*, **94**, 306–339.

Stiner, M. C., Munro, N. D. & Surovell, T. A. (2000). The tortoise and the hare. Small-game use, the broad spectrum revolution, and paleolithic demography. *Current Anthropology*, **41**, 39–74.

Stiner, M. C., Munro, N. D., Surovell, T. A., Tchernov, E. & Bar-Yosef, O. (1999). Pale-olithic population growth pulses evidenced by small animal exploitation. *Science*, **283**, 190–194.

Stoneking, M., Fontius, J. J., Clifford, S. L. *et al.* (1997). *Alu* insertion polymorphisms and human evolution: evidence for a larger population size in Africa. *Genome Research*, **7**, 1061–1071.

Stopp, M. (2002). Ethnohistoric analogues for storage as an adaptive strategy in north-eastern subarctic prehistory. *Journal of Anthropological Archaeology*, **21**, 301–328.

Stout, D. (2002). Skill and cognition in stone tool production. An ethnographic case study from Irian Jaya. *Current Anthropology*, **43**, 693–722.

Straus, L. G. (2000). A quarter-century of research on the Solutrean of Vasco-Cantabria, Iberia and beyond. *Journal of Anthropological Research*, **56**, 39–58.

Straus, L. G. (2001). Africa and Iberia in the Pleistocene. *Quaternary International*, **75**, 91–102.

Straus, L. G. & Bar-Yosef, O. (2001). Out of Africa in the Pleistocene: an introduction. *Quaternary International*, **75**, 1–3.

Straus, L. G., Bicho, N. & Winegardner, A. C. (2000). The Upper Palaeolithic settlement of Iberia: first-generation maps. *Antiquity*, **74**, 553–566.

Straus, L. G. & Winegardner, A. C. (2000). The Upper Paleolithic Settlement of Spain.

In *Paleolítico da Península Ibérica*, Actas do 3. o Congresso de Arqueologia Peninsular, **2**, 443–456.

Strauss, E. (1999). Can Mitochondrial Clocks Keep Time? *Science*, **283**, 1435–1438.

Street, M. & Terberger, T. (1999). The last Pleniglacial and the human settlement of central Europe: new information from the Rhineland. *Antiquity*, **73**, 259–272.

Stringer, C. B. (1989). The origin of modern humans: A comparison of the European and non-European evidence. In *The Human Revolution*, ed. P. Mellars & C. B. Stringer, pp. 232–244. Edinburgh, Edinburgh University Press.

Stringer, C. (2002a). Modern human origins: progress and prospects. *Philosophical Transactions of the Royal Society, London, Series B*, **357**, 563–579.

Stringer, C. (2002b). The morphological and behavioural origins of modern humans. In *The Speciation of Modern* Homo sapiens, ed. T. J. Crow, pp. 23–30. Oxford, Oxford University Press.

Stringer, C. B. & Andrews, P. (1988). Genetic and fossil evidence for the origin of modern humans. *Science*, **239**, 1263–1268.

Stringer, C., Barton, R. N. E. & Finlayson, C. (2000). *Neanderthals on the Edge*. Oxford, Oxbow Books.

Stringer, C. & Gamble, C. (1993). *In Search of the Neanderthals*. London, Thames & Hudson.

Strkalj, G. (2000). The conflict of pre-paradigm schools in modern human origins research. *Acta Biotheoretica*, **48**, 65–71.

Stuiver, M. & Grootes, P. M. (2000). GISP2 Oxygen Isotope Ratios. *Quaternary Research*, **53**, 277–284.

Suc, J. P., Drivaliari, A., Bessais, E. *et al.* (1994). Mediterranean Pliocene vegetation and climate: how to quantify the climate parameters? *US Geological Survey Open-File Report*, 94–23(10).

Suc, J. P. & Zagwijn, W. H. (1983). Plio-Pleistocene correlations between the Northwestern Mediterranean region and Northwestern Europe according to recent biostratigraphic and paleoclimatic data. *Boreas*, **12**, 153–166.

Sugiyama, M. S. (2001). Food, foragers, and folklore: the role of narrative in human subsistence. *Evolution and Human Behaviour*, **22**, 221–240.

Swezey, C. (2001). Eolian sediment responses to late Quaternary climate changes: temporal and spatial patterns in the Sahara. *Palaeogeography, Palaeoclimatology, Palaeoecology*, **167**, 119–155.

Swisher III, C. C., Rink, W. J., Anton, S. C. *et al.* (1996). Latest *Homo erectus* of Java: potential contemporaneity with *Homo sapiens* in southeast Asia. *Science*, **274**, 1870–1874.

Szabó, G. & Hauert, C. (2002). Phase transitions and volunteering in spatial public goods games. *Physical Review Letters*, **89**, 118101-1-118101–4.

Taberlet, P., Fumagalli, L., Wust-Saucy, A-G. & Cosson, J-F. (1998). Comparative phylogeography and postglacial colonization routes in Europe. *Molecular Ecology*, **7**, 453–464.

Takahata, N. & Satta, Y. (1997). Evolution of the primate lineage leading to modern humans: phylogenetic and demographic inferences from DNA sequences. *Proceedings of the National Academy of Sciences, USA*, **94**, 4811–4815.

Taneyhill, D. E. (2000). Metapopulation dynamics of multiple species: the geometry of competition in a fragmented habitat. *Ecological Monographs*, **70**, 495–516.

Tattersall, I. (2000). Paleoanthropology: the last half-century. *Evolutionary Anthropology*, **9**, 2–16.

Tattersall, I. (2002). The case for saltational events in human evolution. In *The Speciation of Modern* Homo sapiens, ed. T. J. Crow, pp. 49–60. Oxford, Oxford University Press.

Tattersall, I. & Schwartz, J. (1999). Hominids and hybrids. The place of Neanderthals in human evolution. *Proceedings of the National Academy of Sciences, USA*, **96**, 7117–7119.

Tattersall, I. & Schwartz, J. (2000). *Extinct Humans*. New York, Westview Press.

Tchernov, E. (1992). Biochronology, paleoecology, and dispersal events of hominids in the southern Levant. In *The Evolution and Dispersal of Modern Humans in Asia*, ed. T. Akazawa, K. Aoki & T. Kimura, 149–188. Tokyo, Hokusen-sha.

Tchernov, E. (1998). The faunal sequence of the southwest Asian Middle Palaeolithic in relation to hominid dispersal events. In *Neandertals and Modern Humans in western Asia*, ed. T. Akazawa, K. Aoki & O. Bar-Yosef, pp. 77–90. New York, Plenum Press.

Templeton, A. (2002). Out of Africa again and again. *Nature*, **416**, 45–51.

Thieme, H. (1997). Lower Palaeolithic hunting spears from Germany. *Nature*, **385**, 807–810.

Thompson, L. G., Yao, T., Davis, M. E. *et al.* (1997). Tropical climate instability: the last glacial cycle from a Qinghai-Tibetan ice core. *Science*, **276**, 1821–1825.

Thorne, A., Grun, R., Mortimer, G., Spooner, N. A. *et al.* (1999). Australia's oldest human remains: age of the Lake Mungo 3 skeleton. *Journal of Human Evolution*, **36**, 591–612.

Tilman, D. & Karieva, P. (Ed.) (1997). *Spatial Ecology. The Role of Space in Population Dynamics and Interspecific Interactions*. Princeton, Princeton University Press.

Tishkoff, S. A., Dietzsch E., Speed, W. *et al.* (1996). Global patterns of linkage disequilibrium at the CD4 locus and modern human origins. *Science*, **271**, 1380–1387.

Tobien, H. (1996). Evolution of zygodons with emphasis on dentition. In *The Proboscidea. Evolution and Palaeoecology of Elephants and Their Relatives*, ed. J. Shoshani & P. Tassy, pp. 76–85. Oxford, Oxford University Press.

Todd, N. E. & Roth, V. L. (1996). Origin and radiation of the Elephantidae. In *The Proboscidea. Evolution and Palaeoecology of Elephants and Their Relatives*, ed. J. Shoshani & P. Tassy, pp. 193–202. Oxford, Oxford University Press.

Tomasello, M. (1999). The human adaptation for culture. *Annual Review of Anthropology*, **28**, 509–529.

Torroni, A., Bandelt, H-J., D'Urbano, L. (1998). MtDNA Analysis reveals a major late Palaeolithic population expansion from southwestern to northeastern Europe. *American Journal of Human Genetics*, **62**, 1137–1152.

Torroni, A., Bandelt, H-J., Macaulay, V. *et al.* (2001). A signal, from human mtDNA, of postglacial recolonization in Europe. *American Journal of Human Genetics*, **69**, 844–852.

Tougard, C. (2001). Biogeography and migration routes of large mammal faunas in South-East Asia during the Late Middle Pleistocene: focus on the fossil and extant faunas from Thailand. *Palaeogeography, Palaeoclimatology, Palaeoecology*, **168**, 337–358.

Tregenza, T. & Butlin, R. K. (1999). Speciation without isolation. *Nature*, **400**, 311–312.

Trinkaus, E. (1981). Neanderthal limb proportions and cold adaptation. In *Aspects of human evolution*, ed. C. B. Stringer, pp. 187–224. London, Taylor & Francis.

Trinkaus, E. (1984). Neanderthal pubic morphology and gestation length. *Current Anthropology*, **25**, 508–514.

Trinkaus, E. (1986). Neanderthals and modern human origins. *Annual Review of Anthropology*, **15**, 193–218.

Trinkaus, E. (1992). Morphological contrasts between Near Eastern Qafzeh-Skhul and late archaic human samples: grounds for a behavioral difference? In *The Evolution and Dispersal of Modern Humans in Asia*, ed. T. Akazawa, K. Aoki & T. Kimura, pp. 277–294. Tokyo, Hokusen-sha.

Trinkaus, E. (1993). Femoral neck-shaft angles of the Qafzeh-Skhul early modern humans, and activity levels among immature Near Eastern Middle Palaeolithic hominids. *Journal of Human Evolution*, **25**, 393–416.

Trinkaus, E. (1997). Appendicular robusticity and the paleobiology of modern human emergence. *Proceedings of the National Academy of Sciences, USA*, **94**, 13367–13373.

Trinkaus, E. & Rhoads, M. L. (1999). Neandertal knees: power lifters in the Pleistocene? *Journal of Human Evolution*, **37**, 833–859.

Trivers, R. L. (1971). The evolution of reciprocal altruism. *Quarterly Review of Biology*, **46**, 35–57.

Turner, C. (2002). Formal status and vegetational development of the Eemian interglacial in northwestern and southern Europe. *Quaternary Research*, **58**, 41–44.

Turner, M. G., Gardner, R. H. & O'Neill, R. V. (1995). Ecological dynamics at broad scales: ecosystems and landscapes. *Science Biodiversity Policy, Bioscience, 5* (Supplement), 29–35.

Tzedakis, P. C. (1993). Long-term tree populations in northwest Greece through multiple Quaternary climatic cycles. *Nature*, **364**, 437–440.

Tzedakis, P. C. (1994). Vegetation change through glacial-interglacial cycles: a long pollen sequence perspective. *Philosophical Transactions of the Royal Society, London, Series B.*, **345**, 403–432.

Tzedakis, P. C., Frogley, M. R. & Heaton, T. H. E. (2002). Duration of last interglacial conditions in northwestern Greece. *Quaternary Research*, **58**, 53–55.

Tzedakis, P. C., Lawson, I. T., Frogley, M. R., Hewitt, G. M. & Preece, R. C. (2002). Buffered tree population changes in a Quaternary refugium: evolutionary implications. *Science*, **297**, 2044–2047.

Ukkonen, P., Lunkka, J. P., Jungner, H. & Donner, J. (1999). New radiocarbon dates from Finnish mammoths indicating large ice-free areas in Fennoscandia during the Middle Weichselian. *Journal of Quaternary Science*, **14**, 711–714.

Ulijaszek, S. J. (2001). Potential seasonal ecological challenge of heat strain among Australian Aboriginal people practicing traditional subsistence methods: a computer simulation. *American Journal of Physical Anthropology*, **116**, 236–245.

Underhill, P. A., Shen, P., Lin, A. A. *et al.* (2000). Y chromosome sequence variation and the history of human populations. *Nature Genetics* **26**, 358–361.

van Andel, T. H. (1998). Middle and Upper Palaeolithic environments and ^{14}C dates beyond 10,000 BP. *Antiquity*, **72**, 26–33.

van Andel, T. H. & Tzedakis, P. C. (1996). Palaeolithic landscapes of Europe and environs, 150,000–25,000 years ago. *Quaternary Science Reviews*, **15**, 481–500.

van Andel, T. H. & Tzedakis, P. C. (1998). Priority and opportunity: reconstructing the European Middle Palaeolithic climate and landscape. In *Science in Archaeology: An Agenda for the Future*, ed. J. Bayley. pp. 37–46. London, English Heritage.

Vandenberghe, J., Coope, R. & Kasse, K. (1998). Quantitative reconstructions of palaeoclimates during the last interglacial-glacial in western and central Europe: an introduction. *Journal of Quaternary Science*, **13**, 361–366.

Van Valkenburgh, B. (1994). Ecomorphological analysis of fossil vertebrates and their palaeocommunities. In *Ecological Morphology. Integrative Organismal Biology*, ed. P. C. Wainwright & S. M. Reilly. pp. 140–166. Chicago, University of Chicago Press.

van der Made, J. (1999). Ungulates from Atapuerca TD6. *Journal of Human Evolution*, **37**, 389–414.

van Peer, P. (1998). The Nile corridor and the Out-of-Africa model. *Current Anthropology*, **39** (Suppl.), 5515–5140.

van Schaik, C. P., Ancrenaz, M., Borgen, G. *et al.* (2003). Orangutan cultures and the evolution of material culture. *Science*, **299**, 102–105.

Vartanyan, S. L., Garutt, V. E. & Sher, A. V. (1993). Holocene dwarf mammoths from Wrangel Island in the Siberian Arctic. *Nature*, **362**, 337–340.

Vega-Toscano, L. G. (1990). Le fin du paléolithique moyen au sud de l'Espagne: ses implications dans le contexte de la Péninsule Ibérique. *Memoires du Musée de Prehistoire d'Ille-de-France*, **3**, 169–176.

Vega-Toscano, L. G., Hoyos, M., Ruiz Bustos, M. & Laville, H. (1988). La sequence de la grotte de la Carihuela (Piñar, Grenade): Chronostratigraphie et paléoécologie du Pléistocene Supérieur du Sud de la Péninsule Ibérique. In *L'homme de Néandertal*, vol.2, *L'environnement*, ed. M. Otte. pp. 169–180. Liege, Eraul.

Verschuren, D., Laird, K. R. & Cumming, B. F. (2000). Rainfall and drought in equatorial east Africa during the past 1,100 years. *Nature*, **403**, 410–414.

Vigilant, L., Stoneking, M., Harpending, H., Hawkes, K. & Wilson, A. C. (1991). African populations and the evolution of human mitochondrial DNA. *Science*, **253**, 1503–1507.

Villaverde, V. & Fullola, J. Ma. (1990). Le Solutréen de la zone mediterranéenne espagnole. Feuilles de pierre. Les industries à pointes foliacées du Paléolithique supérieur européen, Kraków 1989. *Eraul*, **42**, 467–480.

Voelker, A. H. L. (2002). Global distribution of centennial-scale records for Marine Isotope Stage (MIS) 3: a database. *Quaternary Science Reviews*, **21**, 1185–1212.

Walker, A. C. (1981). Dietary hypotheses in human evolution. *Philosophical Transactions of the Royal Society, London, Series B*, **292**, 47–64.

Wang, Z-L., Wang, F-Z., Chen, S. & Zhu, M-Y. (2002). Competition and coexistence in regional habitats. *American Naturalist*, **159**, 498–508.

Wang, Z-L., Zhang, J-G. & Liu, X-M. (2000). Competition and coexistence in spatially subdivided habitats. *Journal of Theoretical Biology*, **205**, 631–639.

Watts, W. A., Allen, J. R. M. & Huntley, B. (2000). Palaeoecology of three interstadial events during oxygen-isotope Stages 3 and 4: a lacustrine record from Lago Grande di Monticchio, southern Italy. *Palaeogeography, Palaeoclimatology, Palaeoecology*, **155**, 83–93.

Webb, R. S., Rind, D. H., Lehman, S. J., Healy, R. J. & Sigman, D. (1997). Influence of ocean heat transport on the climate of the Last Glacial Maximum. *Nature*, **385**, 695–699.

Webb, T., III & Bartlein, P. J. (1992). Global changes during the last 3 million years: climatic controls and biotic responses. *Annual Review of Ecology and Systematics*, **23**, 141–173.

Weber, A. W., Link, D. W. & Katzenberg, M. A. (2002). Hunter–gatherer culture change and continuity in the Middle Holocene of the Cis-Baikal, Siberia. *Journal of Anthropological Archaeology*, **21**, 230–299.

Wedekind, C. & Milinski, M. (2000). Cooperation through image scoring in humans. *Science*, **288**, 850–852.

Weiss, E. (2002). Drought-related changes in two hunter–gatherer California populations. *Quaternary Research*, **58**, 393–396.

Wells, R. S., Yuldasheva, N., Ruzibakiev, R. *et al.* (2001). The Eurasian heartland: a continental perspective on Y-chromosome diversity. *Proceedings of the National Academy of Sciences, USA*, **98**, 10244–10249.

Wheeler, P. E. (1994). The thermoregulatory advantages of heat storage and shade-seeking behaviour to hominids foraging in equatorial savannah environments. *Journal of Human Evolution*, **26**, 339–350.

Wheeler, P. E. (1996). The environmental context of functional body hair loss in hominids (a reply to Amaral, 1996). *Journal of Human Evolution*, **30**, 367–371.

White, R. (1985). *Upper Palaeolithic land use in the Périgord: A topographic approach to subsistence and settlement.* Oxford, British Archaeological Reports Series No. 253.

White, T. D. & Harris, J. M. (1977). Suid evolution and correlation of African hominid localities. *Science*, **198**, 13–21.

Wiens, J. A. (1997). Metapopulation dynamics and landscape ecology. In *Metapopulation Biology. Ecology, Genetics, and Evolution*, ed. I. A. Hanski & M. E. Gilpin, pp. 43–62. San Diego, Academic Press.

Willis, K. J. (1996). Where did all the flowers go? The fate of temperate European flora during glacial periods. *Endeavour* **20**, 110–114.

Willis, K. J., Rudner, E. & Sumegi, P. (2000). The full-glacial forests of central and southeastern Europe. *Quaternary Research*, **53**, 203–213.

Willis, K. J., Rudner, E. & Sumegi, P. (2001). Reply to Carcaillet and Vernet. *Quaternary Research*, **55**, 388–389.

Willis, K. J. & Whittaker, R. J. (2000). The refugial debate. *Science*, **287**, 1406–1407.

Wilson, A. B., Noack-Kunnmann, K. & Meyer, A. (2000). Incipient speciation in sympatric Nicaraguan crater lake cichlid fishes: sexual selection versus ecological diversification. *Proceedings of the Royal Society London, Series B*, **267**, 2133–2141.

Wilson, J. (1975). The last glacial environment at the Abri Pataud: a possible comparison. In *Excavation of the Abri Pataud, Les Eyzies (Dordogne)*, ed. H. L. Movius, Jnr, pp. 175–186. Cambridge, MA, Peabody Museum.

Woillard, G. M. (1978). Grand Pile peat bog: a continuous pollen record for the last 140,000 years. *Quaternary Research*, **9**, 1–21.

Woillard, G. (1979). Abrupt end of the last interglacial in north-east France. *Nature*, **281**, 558–562.

Woillard, G. M. & Mook, W. G. (1982). Carbon-14 dates at Grande Pile: correlation of land and sea chronologies. *Science*, **215**, 159–161.

Wolpoff, M. H. (1989). Multiregional evolution: the fossil alternative to Eden. In *The Human Revolution. Behavioural and Biological Perspectives in the Origins of Modern Humans*, ed. P. Mellars & C. Stringer, pp. 62–108. Edinburgh, Edinburgh University Press.

Wolpoff, M. (1998). Concocting a divisive theory. *Evolutionary Anthropology*, **7**, 1–3.

Wolpoff, M. H. & Caspari, R. (2000). The many species of humanity. *Anthropological Review*, **63**, 3–17.

Wood, B. & Collard, M. (1999). The human genus. *Science*, **284**, 65–71.

Wooller, M. J., Swain, D. L., Ficken, K. J., Agnew, A. D. Q., Street-Perrott, F. A. & Eglington, G. (2003). Late Quaternary vegetation changes around Lake Rutundu, Mount Kenya, East Africa: evidence from grass cuticles, pollen and stable carbon isotopes. *Journal of Quaternary Science*, **18**, 3–15.

Wright, A. A., Rivera, J. J. & Shyan, M. (2000). Music perception and octave generalization in rhesus monkeys. *Journal of Experimental Psychology*, **129**, 291–307.

Yellen, J. E., Brooks, A. S., Cornelissen, E., Mehlman, M. J. & Stewart, K. (1995). A Middle Stone Age worked bone industry from Katanda, Upper Semliki Valley, Zaire. *Science*, **268**, 553–556.

Zabel, M., Schneider, R. R., Wagner, T., Adegbie, A. T., de Vries, U. & Kolonic, S. (2001). Late Quaternary climate changes in central Africa as inferred from terrigenous input to the Niger Fan. *Quaternary Research*, **56**, 207–217.

Zagwijn, W. H. (1992). Migration of vegetation during the Quaternary in Europe. *Courier Forschungs-Institut Senckenberg*, **153**, 9–20.

Zeder, M. A. & Hesse, B. (2000). The initial domestication of goats (*Capra hircus*) in the Zagros Mountains 10,000 years ago. *Science*, **287**, 2254–2257.

Zilhao, J. (1993). Le passage du Paléolithique moyen au Paléolithique supérieur dans le Portugal. In *El origen del hombre moderno en el Suroeste de Europa*, ed. V. Cabrera Valdes, pp. 127–125. Madrid, UNED.

Zilhao, J. (1995). *O Paleolitico Superior da Estremadura Portuguesa*. PhD Thesis, Lisbon, University of Lisbon.

Zilhao, J. (1996). The extinction of Iberian Neanderthals and its implications for the origins of Modern Humans in Europe. *Actes du XIII Congres International des Sciences Préhistoriques et Protohistoriques*, **2**, 299–312. Forli, Abaco.

Zilhao, J. (2001). Radiocarbon evidence for maritime pioneer colonization at the origins of farming in west Mediterranean Europe. *Proceedings of the National Academy of Sciences, USA*, **98**, 14180–14185.

Zilhao, J. & d'Errico, F. (1999). The chronology and taphonomy of the earliest Aurignacian and its implications for the understanding of Neanderthal extinction. *Journal of World Prehistory*, **13**, 1–68.

Zilhao, J. & d'Errico, F. (2000). La nouvelle 'bataille aurignacienne'. Une revision critique de la chronologie de Chatelperronien et de l'Aurignacien ancien. *L'Anthropologie*, **104**, 17–50.

Zilhao, J. & Trinkaus, E. (Eds) (2002). *Portrait of the Artist as a Child. The Gravettian Human Skeleton from the Abrigo do Lagar Velho and its Archeological Context.* Trabalhos de Arqueologia 22. Lisboa, IPA.

Zubrow, E. (1989). The demographic modelling of Neanderthal extinction. In *The Human Revolution. Behavioural and Biological Perspectives in the Origins of Modern Humans*, ed. C. Stringer & P. Mellars, pp. 212–231. Edinburgh, Edinburgh University Press.

Index